SpringerBriefs in Phys

For further volumes:
http://www.springer.com/series/8902

Yasuhito Narita

Plasma Turbulence
in the Solar System

 Springer

Yasuhito Narita
Institut für Geophysik und
    extraterrestrische Physik
Technische Universität Braunschweig
Mendelssohnstraße 3
D-38106 Braunschweig
Germany
e-mail: y.narita@tu-bs.de

ISSN 2191-5423   e-ISSN 2191-5431
ISBN 978-3-642-25666-0   e-ISBN 978-3-642-25667-7
DOI 10.1007/978-3-642-25667-7
Springer Heidelberg Dordrecht London New York

Library of Congress Control Number: 2011943321

Printed on acid-free paper

Springer is part of Springer Science+Business Media (www.springer.com)

Preface

Turbulence represents random motions of flow, and is one of the common experiences in our daily life as can be seen in air and water flow. Whether or not turbulence exists in the extraterrestrial world turns out to be one of the fundamental questions in understanding astrophysical systems. Stellar dynamics, interplanetary and interstellar space, cosmic ray, and accretion disks - these systems are largely maintained by the existence of turbulence, but their dynamical behaviors are substantially different from that of ordinary gas or fluid, since the medium is an ionized gas, called the plasma, and is electrically conducting

Plasma turbulence is a challenge in physics both in theories and observations, and the aim of this book is to review plasma turbulence on the introductory level and also to review recent developments and knowledge obtained by Cluster, the multi-spacecraft mission. Cluster is a four-spacecraft mission in near-Earth space and has enabled us for the first time to determine spatial structures of space plasma dynamics. Different branches of physics are involved in plasma turbulence: fluid dynamics, electromagnetism and electrodynamics, plasma physics, geophysics, space, and astrophysics. Studying plasma turbulence requires these backgrounds, while progress in plasma turbulence research has mutual impacts on these subjects in return, bringing the different research fields together.

This book is organized in the following fashion. Chapter 1 introduces the concept of plasma turbulence with its historical developments. Chapter 2 is a review of theoretical building blocks for understanding plasma turbulence. Chapter 3 presents the analysis methods for Cluster data. Chapter 4 is a review of plasma turbulence studies using Cluster data in near-Earth space. Chapter 5 presents the impacts of plasma turbulence on the related subjects: plasma turbulence as general physics problem, as astrophysics problem, and as Earth science problem.

Space plasma research has entered a new era with multi-spacecraft missions. It is a pleasure to the author if students and researchers in other fields become interested in physics in the extraterrestrial world, and use this book as a guide to this subject.

The writing of this book was performed under the auspices of Institut für Geophysik und extraterrestrische Physik, Technische Universität Braunschweig as well as Kavli Institute for Theoretical Physics. The author thanks all colleagues in theory and experiment who, through their unselfish collaborations with the author, have substantially contributed to the insights and interpretations found in this book. Among these many individuals I would like to thank Karl-Heinz Glassmeier and David Gross who have been particularly generous, cooperative, and persistent in supporting the author's efforts to understand the subject of the book. The research described here has been supported by Budesministerium fürWirtschaft und Technologie and Deutsches Zentrum für Luft- und Raumfahrt under contract 50 OC 0901, and in part by the National Science Foundation under Grant No. NSF PHY05-51164.

Braunschweig   Yasuhito Narita

Contents

Chapter 1
Introduction

Abstract Plasma turbulence in the solar system is an interdisciplinary field of study and also a challenge in physics in theory and observation. This chapter reviews historical developments of the two distinct subjects: turbulence research and space plasma research, and introduces fundamental concepts and notions such as energy cascade in turbulence as well as the plasma physical picture of the solar system.

1.1 What is Plasma Turbulence?

Solar system science is one of interdisciplinary research fields, and has developed together with space technology: rockets, spacecraft, space telescopes, and so on. It is the world where plasma physics plays a major role. The reason for this is the fact that the gas is so hot and dilute that it remains in the plasma state as ionized gas. Plasmas exhibit various kinds of dynamics, and in particular it behaves as a turbulent fluid under some conditions. Plasma turbulence plays an important role not only in many places in the solar system such as at the Sun, in the interplanetary space, and in planetary magnetospheres, but also in many astrophysical systems such as star and planet formation, accretion disks, interstellar medium, and cosmic ray.

Plasma turbulence is the phenomenon that plasma and electromagnetic field fluctuate randomly in appearance. It is one of major challenges in physics both in the theoretical and observational senses. The solar system serves as the only accessible, natural laboratory of astrophysical plasmas as it is a collisionless gas (Sect. 2.1.1.2) and the scale of dynamics cannot be achieved in laboratory experiments on the ground. Studying plasma turbulence in the solar system necessarily involves fundamental physics (plasma physics, fluid dynamics, electromagnetism and electrodynamics), solar system science (the Sun and its interaction with the Earth and planets), and signal processing of spacecraft data. The study has in return impacts on these subjects. Knowledge of plasma turbulence is valuable in understanding the problems of space weather and possibly Earth climate, too.

Y. Narita, *Plasma Turbulence in the Solar System*, SpringerBriefs in Physics, DOI: 10.1007/978-3-642-25667-7_1, © The Author(s) 2012

Fig. 1.1 Water falling into a
pool, sketched by Leonardo
da Vinci in ca. 1508–1513

1.2 Historical Development

Turbulence can be found in many kinds of fluids such as air or water flow. Its existence
and importance in the extrasolar world such as in the solar system or astrophysical
systems have been revealed together with the progress in space plasma research.
Plasma turbulence in the solar system exhibits both plasma physical and fluid turbulence characters. Some processes are similar between fluid turbulence and plasma
turbulence, while others are different. These two research fields have different backgrounds and their historical developments are reviewed.

1.2.1 View from Turbulence Research

Recognition of turbulence as a physics problem dates back to Leonardo Da Vinci.
His sketch in the early sixteenth century (Fig. 1.1) is the first scientific recognition
and description of fluid turbulence. It was noticed that turbulence exhibits eddies
at different sizes and in different directions. He also noted in the sketch: "Observe
the motion of the surface of the water, which resembles that of hair, which has two
motions, of which one is caused by the weight of the hair, the other by the direction
of the curls; thus the water has eddying motions, one part of which is due to the
principal current, the other to random and reverse motion" [4].

Fluid dynamics was then established through contributions by celebrated math-
ematicians and physicists such as Euler, Navier, Stokes, Taylor, and so on, from
seventeenth to twentieth century. Even to date turbulence is recognized as one of the
unsolved problems in physics and mathematics. For example, it is not known if the
Navier–Stokes equation has a unique solution. Only few special solutions are known
in fluid dynamics.

Current understanding of fluid turbulence owes a lot to the picture of energy
cascade proposed by Richardson and Kolmogorov. Figure 1.2 is an illustration of
this concept. Large-scale eddies are created by an external force (referred to as
the energy injection), and energy is put into the flow system. A certain instability
process is operating which deforms the large-scale eddies and makes the eddies split
into those on a smaller scale. The instability then acts on that small-scale eddies,

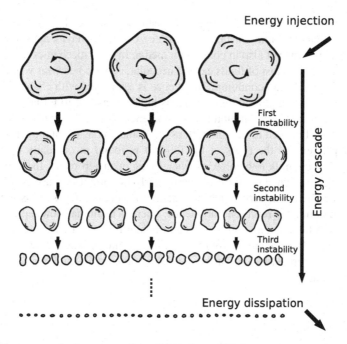

Fig. 1.2 Energy cascade picture according to Richardson and Kolmogorov. Illustration after Frisch, Turbulence (1995), Cambridge University Press [3]

creating eddies at an even smaller size. The instability operates successively and eddies split into ever smaller sizes (energy cascade), until finally the smallest scale is reached and eddies disappear due to the viscosity of the fluid (energy dissipation). Fluid turbulence thus consists of eddies at various sizes.

The concept of chaos needs to be addressed here. The governing equation, the Navier–Stokes equation, is in fact known and it is deterministic. In other words, we know what kind of forces are acting on the fluid elements and it should be possible to calculate the future behavior of the flow using the equation. Nevertheless, predicting the future behavior of the flow is very difficult, and this is due to the presence of nonlinearity. It is the effect that makes the fluid motion so random and irregular that even smallest difference or uncertainty at the initial condition ends up with completely different states. This problem, called the chaos, was recognized even earlier before the development of quantum mechanics which states that particle position and momentum cannot be measured and determined with arbitrary accuracy due to Heisenberg's uncertainty principle, and instead, one can only study the probability of particle position and momentum. It should be noted, however, that in quantum mechanics this probability itself is predictable. In turbulence, in a similar fashion, predicting the state of individual fluid elements is virtually impossible because of nonlinearity. Therefore the emphasis of turbulence studies is on its statistical properties, i.e. we are not studying the exact future state of the flow. This situation is reminiscent of the difference between the concept of climate (statistical property) and that of weather (instantaneous state). Chaotic behavior appears not only in dynamical systems but also in continuum media. In fluid mechanics the equation is also nonlinear and most of fluids exhibit irregular, chaotic behavior when it is set to motion under a certain condition. Turbulence is associated with chaotic behavior of the fluid motion and the future behavior is highly sensitive to initial conditions; prediction of the exact fluid motion under turbulence is virtually impossible.

Turbulence is believed to be a ubiquitous phenomenon, as it is a property of flow and is independent on the types of the fluid. It is found in our daily experience when pouring some milk into coffee or when flying in an aeroplane. In engineering, understanding turbulence properties is particularly important because one needs to regulate the occurrence of turbulence in modeling cars or aeroplanes for its aero-drag.

Not all fluid motions represent turbulence. There is a laminar flow representing a regular, smooth flow pattern and there is a turbulent flow representing an irregular, random pattern. The transition from the laminar to the turbulent flow can conveniently be characterized by one of the control parameters in fluid dynamics called the Reynolds number,

$$ \mathrm{Re} = \frac{UL}{\nu}, \tag{1.1} $$

where U, L, and ν denote the flow speed, the length scale of the flow, and the viscosity. Flow around a cylinder is displayed as an example of transition from a laminar to a turbulent flow in Chap. 2. Coffee-milk problem is another example of transition into turbulence. When milk is very gently poured into coffee, the milk stays at the bottom of the cup. While there is certainly a molecular diffusion process that mixes the milk and the coffee into a homogeneous state, this process is usually slow and it takes hours to days for the two fluids to be mixed. But when we stir the coffee and the milk with a spoon, the two fluids are immediately mixed. Turbulence provides a means of effective mixing.

Another breakthrough after the physical picture of turbulence was given by Richardson and Kolmogorov was made by Kolmogorov himself, Monin and Yaglom, and Kraichnan. In particular, Kraichnan proposed to solve the Navier–Stokes equation in a statistical treatment, known as the Direct Interaction Approximation (DIA), and indeed his theory justifies the Richardson–Kolmogorov picture of energy cascade without introducing any arbitrary, free parameter. To date, various models and methods are known and used in a variety of applications from physics to engineering.

1.2.2 View from Space Plasma Research

The importance of electric currents and electromagnetic field in space was already recognized by Birkeland in the early twentieth century. In 1940s the notion of treating plasmas as electrically conducting fluid (or magnetized, conducting fluid) was proposed by Alfvén. This subject is nowadays known as magnetohydrodynamics or MHD. It is a fluid picture of plasma, and conveniently describes the motion of plasma using macroscopic quantities such as density, bulk velocity, and temperature. On the other hand, MHD is a set of nonlinear equations (nonlinearity in gas dynamics and in coupling between gas and electromagnetic field) that leads to turbulence phenomena in plasma. Plasmas physics in the solar system has developed as an extension of atmospheric research to the ionosphere and further to the magnetosphere, and is of interdisciplinary nature from the era of earlier studies. The subject of space plasma also represents a crossroad between geophysics and astrophysics. Various kinds of observations have been devised and used to diagnose plasma dynamics in the Sun, in the interplanetary space, and in Earth's and planetary magnetospheres.

Ground-based observations have been performed using telescopes and magnetometers over decades to centuries. The scientific use of telescope dates back to the early seventeenth century when Galilei made pioneering works with sunspots and Jovian satellite observations. The use of magnetometer began in 1830s when Gauss and Weber invented the first magnetometer measuring the strength of Earth's magnetic field. In tweentieth century ground-based observations made extensive use of telescopes, magnetometer, and radio antenna. Radio wave observations provided the first evidence of Jupiter's magnetic field; Energetic particles from galaxy were found, known as cosmic rays nowadays; The existence of the magnetic field in the Sun (sunspots, strictly speaking) was found by Hale in the early tweentieth century using Zeeman effect in the spectroscopic measurement. Space plasmas have been studied in situ in Earth's ionosphere and magnetosphere using rockets and spacecraft since 1950s. The plasma and electromagnetic field can be measured along the trajectory of spacecraft in various places in the interplanetary space and in the planetary environments.

1.2.2.1 Sun and Heliosphere

The Sun is one of typical, middle-aged, main-sequence stars in the Universe, and also represents a spherical object consisting of plasma, mostly protons and electrons. It is also a factory of chemical elements. Gravity in the core region of the Sun is so strong and the gas temperature is so high that nuclear fusion process is operating. The elements such as helium, oxygen, and carbon are created from the fusion reaction. This process releases sufficient energy (mostly as gamma rays and thermal energy), which is converted into different types of energy. The produced energy is used for further purposes: further nuclear reaction producing chemical elements and other elementary particles (neutrinos), radiation or electromagnetic fields at various wavelengths, heat, and gas convection and dynamo mechanism generating the large-scale magnetic field.

The interior structure of the Sun is divided into three distinct zones: core, radiative zone, and convective zone. In the core, energy is released by the hydrogen-fusion reaction. In the radiative zone, the produced energy is transported from the core to the outer layer of the solar interior by radiation process (absorption and re-emission of photons). The outer layer is referred to as the convective zone in which gas is circulating primarily in the radial direction by convection and the energy is transported to the solar surface. Granular structure at the surface represents the convection cell. The energy transported by radiation and convection is released at the surface in the form of blackbody radiation with the effective temperature at about 5780 K. The Sun as a whole is in a hydrostatic equilibrium, a balance between inward contraction due to self-gravity and outward expansion due to high thermal pressure.

The Sun has atmospheres in layer. The inner layer, the intermediate layer, and the outer layer are called the photosphere, the chromosphere, and the corona, respectively. Coronal gas temperature reaches 1 million Kelvin, and the question why the gas temperature rises abruptly from 5780 K at the photosphere to 10^6 K in corona is one of the unsolved problems in solar physics. Since corona is so hot, i.e. thermal pressure is so high the gas is in the fully ionized state, the plasma. The coronal plasma streams radially outward from the Sun and this plasma stream is referred to as the solar wind. The plasma is accelerated and the flow speed exceeds the sound speed. The solar wind is therefore a radially expanding supersonic flow plasma stream.

The gas in the convective zone is also in the plasma state and electrically conducting. The magnetic field is coupled to the plasma, and is amplified by the convective motion, the Coriolis force due to rotation, and the differential rotation (or shear flow). Buoyancy can be very effective in the convective zone and the magnetic flux may rise and pop up at the surface, which is observed as a sunspot.

The solar wind plasma expands in the solar system, carrying the magnetic field. The plasma and the magnetic field in the interplanetary space forms a domain of the solar plasma called the heliosphere (Fig. 1.3) until the solar wind encounters the interstellar medium at distance about 100–150 AU (the distance from the Sun to the Earth is 1 AU, ca. 1.5×10^8 km). The Sun, on the other hand, is rotating around its axis, and the rotation winds the magnetic field lines in the heliosphere. The interplanetary magnetic field has a spiral form. The magnetic field in the heliosphere has an important function for us: it serves as a shield from galactic cosmic ray.

The solar wind is a supersonic flow and is in a fully developed turbulence state. Because of the high conductivity, the flow motion is closely coupled to the electromagnetic field. Solar wind turbulence must be treated as plasma turbulence and understanding its physical processes is the focus of this book. The solar wind plasma is a collisionless gas for the extremely large mean free path. On the other hand, plasmas in general exhibit a variety of excitation modes or wave modes. There are a number of questions on the nature of plasma turbulence, for example, how the energy is transported from one scale to another; and in which direction the energy is transported; what the mechanism of energy dissipation is in a collisionless gas.

1.2.2.2 Planetary Magnetospheres

In situ observations using spacecraft revealed that many planets in the solar system have the magnetic field, forming a spatial domain of the planetary magnetic field called the magnetosphere caused by the interaction with the solar wind: Mercury,

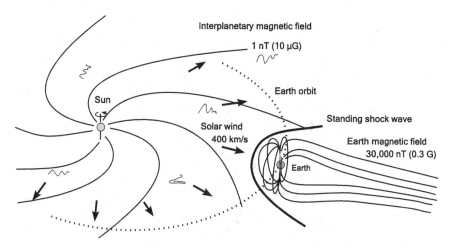

Fig. 1.3 Stereographic view of plasma domains and magnetic field lines in the inner part of heliosphere from the Sun to the Earth

Earth, Jupiter, Saturn, Uranus, and Neptune have the intrinsic magnetic field and therefore magnetospheres. Venus and Mars do not have a global magnetic field, but still a comet-like magnetic field structure is formed around these planets due to the interaction between the solar wind and the planetary ionosphere.

The Earth magnetic field has important function in our lives: it is shielding us from the solar wind which is a supersonic plasma flow, and furthermore from galactic cosmic ray as done by the heliospheric magnetic field. Figure 1.4 displays the meridional cross-section of field lines of the Earth magnetosphere and its surrounding region. On the dayside the solar wind plasma hits the Earth magnetic field and the field has a compressed, dipolar structure. On the nightside the solar wind is passing by the Earth and the magnetic field lines are stretched by the solar wind. Sources of plasma are the Sun and the planet, and electric current flows in various directions to sustain the shape of the planetary magnetosphere. For example, there is a current sheet in the magnetotail, separating the field lines in anti-parallel configuration. The current flows at the boundary layer of the magnetosphere called the magnetopause. Solar wind is a supersonic flow, and a standing shock wave called the bow shock is formed when it encounters an obstacle such as planetary magnetic field or atmosphere. In the case of Earth magnetosphere, the dayside magnetopause is located at distance about 10 R_E (1 Earth radius, R_E, is about 6370 km) from the Earth, and the bow shock at about 15 R_E. Shock wave in space plasma has unique dissipation mechanism: particle reflection. A small portion of incoming, upstream particle population is specularly reflected and streams backward along the magnetic field line against the bulk flow. The shock-upstream region connected to the shock by the magnetic field line is referred to as the foreshock. Both ions and electrons are reflected and they form the respective foreshock regions. Velocity distribution of particles can be a two-beam distribution (the small 3D panel on left in the ion

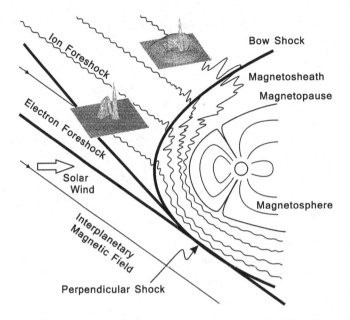

Fig. 1.4 Plasma domains in near-Earth space. Waves are excited around Earth's bow shock with two distinct velocity distributions of ions upstream of the bow shock

foreshock in Fig. 1.4) or a ring-distribution around the solar wind ion beam (the small panel on right). Both shock-upstream region and shock-downstream region (called the magnetosheath) exhibit large-amplitude, random fluctuations of the magnetic field and the plasma, suggesting that the plasma shock distributes the kinetic energy of the plasma flow in the solar wind not only into thermal energy but also into waves and turbulence.

Dynamics of space plasma (solar wind and magnetospheric plasma) has been extensively studied using in situ spacecraft measurements since 1960s, but the observational method was primarily limited to single spacecraft missions, and the studies of waves and turbulence were based on time series data analysis. Separation between temporal variation and spatial variation in three-dimensional space became possible using multi-spacecraft missions such as Cluster [2] and THEMIS [1]. Cluster is a four-spacecraft mission in Earth polar orbit, and performs formation flight with the tetrahedron configuration. THEMIS is a five-spacecraft mission in the Earth's equatorial orbit, and performs one-dimensional array measurement along the same field line in the geomagnetic tail.

References

1. Angelopoulos, V.: The THEMIS Mission. Space Sci. Rev. **141**, 5–34 (2008)
2. Escoubet, C.P., Fehringer, M., Goldstein, M.: The Cluster mission. Ann. Geophys. **19**, 1197–1200 (2001)
3. Frisch, U.: Turbulence, The legacy of A. Cambridge University Press, N. Kolmogorov (1995)
4. Trans. Piomelli in Lumley, J.L.: Some comments on turbulence, Phys. Fluids A. 4:203–211 (1997)

Chapter 2
Theoretical Background

Abstract Elements of plasma physics and fluid turbulence are introduced in this chapter on the introductory level. These are theoretical building blocks for highlighting plasma turbulence in the solar system as well as astrophysical systems as a challenge in physics. Emphasis in plasma physics is on the fluid picture, called magnetohydrodynamics, and similarities and differences are discussed between magnetohydrodynamics and fluid turbulence.

2.1 Elements of Plasma Physics

2.1.1 Origin and Properties of Plasmas

When gas is heated, atoms are dissociated into ions (positively charged) and electrons (negatively charged). The ionized gas is electrically quasi-neutral but conducting due to high mobility of ions and electrons. It exhibits a variety of behaviors and collective motions, and is referred to as the plasma. This name was given by Langmuir in 1928 after the Greek term $\pi\lambda\acute{\alpha}\sigma\mu\alpha$, which means something molded or a moldable substance. Stellar interiors and atmospheres, gaseous nebulae, and the interstellar medium are mostly plasmas. They can be found in near-Earth space (Earth's ionosphere and magnetosphere) and in the solar system (interplanetary space and planetary magnetospheres), too. In our daily life, on the other hand, encounters with plasmas are rather limited: the flash of a lightning, the soft glow of the aurora borealis, the conducting gas inside a fluorescent tube or neon sign.

It was once thought that 99% of matter in the universe was in the plasma state, that is, matter is in the form of ionized gas, since conditions such as high temperature and low density are favorable for maintaining gas ionization. Also, there are sources of ionization other than high temperature. Modern cosmology and astrophysics, however, suggest that there must be a different kind of matter in the Universe other than baryonic or leptonic matters, referred to as the dark matter, in order to

Y. Narita, *Plasma Turbulence in the Solar System*, SpringerBriefs in Physics, DOI: 10.1007/978-3-642-25667-7_2, © The Author(s) 2012

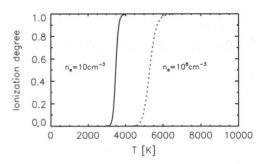

account for radial profile of galaxy rotation rate and distorted images of galaxy
due to gravitational lensing (see recent developments in cosmology, for example,
in [1]). Furthermore, the dark matter must be dominating the total mass density in
the Universe in comparison to the baryonic matter by factor 6 or 7. Yet, the above
statement of the plasma dominance is true for the baryon-lepton matters or "visible"
matters such as protons, neutrons, and electrons. It is interesting that we are living
in the only 1% (or less) of the baryonic Universe in which plasmas do not occur
naturally [2].

2.1.1.1 Ionization Sources

There are a variety of ionization sources in space, but the primary source of ionization
is high temperature. The ratio of the amount of ionization or number density for
ionized atoms (n_{ion}) to that for neutral atoms (n_{atm}) expected in thermal equilibrium
is given by the Saha equation:

$$\frac{n_{\mathrm{ion}}}{n_{\mathrm{atm}}} = \frac{1}{n_{\mathrm{e}}} \left(\frac{2\pi m_{\mathrm{e}} k_{\mathrm{B}} T}{h^2} \right)^{3/2} e^{-U/k_{\mathrm{B}} T}, \tag{2.1}$$

where n_{e} denotes the electron number density, m_{e} the electron rest mass, k_{B} the
Boltzmann constant, T the gas temperature, h the Planck constant, and U the ioniza-
tion potential.

Figure 2.1 displays the profile of ionization degree, $n_{\mathrm{ion}}/(n_{\mathrm{ion}} + n_{\mathrm{atm}})$, as a function of temperature for hydrogen atoms in the ground state (ionization potential
$U = 13.6\,\mathrm{eV}$). The ionization degree is shown for two different gas densities:
$n_{\mathrm{e}} = 10\,\mathrm{cm}^{-3}$ (solid curve) which is typical of the solar wind at the Earth orbit
(1 AU) and $n_{\mathrm{e}} = 10^8\,\mathrm{cm}^{-3}$ (dotted curve) typical of the solar corona at the radius
$1.1 r_\odot$ from the center of the sun (or $7 \times 10^4\,\mathrm{km}$ above the solar surface).

As the temperature is raised, the ionization degree remains low until the ionization
potential is only a few times of the thermal energy, $k_{\mathrm{B}} T$. Then the ionization degree
rises abruptly, and the gas is in the plasma state. The gas then becomes fully ionized.
In astrophysical systems (stars and interstellar medium) gas temperatures are usually
of the order of millions of degrees, while the temperature on the Earth's atmosphere
(on the ground) is about 300 K. Typical values of the electron temperatures in the

solar wind and the corona are 300,000 K and 1,000,000 K, respectively, whereas the jumps in the ionization degree are about 3000 to 3500 K for the solar wind, and 5000 to 6000 K for the corona. The gas can remain fully ionized in the solar wind and corona, but not on the Earth.

The ionization degree depends on the electron density, as well. The ionization temperature becomes lower at lower density, and the slope changes slightly such that it is steeper at lower density and less steep at higher density. The reason for this is that a dilute gas (i.e., lower density) has a longer mean free path and can maintain ionization more easily, and collisions bringing ionized atoms into neutral become more frequent at higher density.

Besides the high temperature, there are other ionization sources. Photoionization is the process in which an incident photon ejects one or more electrons from an atom or molecule, particularly at shorter wavelengths such as UV and also X-ray. Thermal radiation (or blackbody radiation) from stars cover a broad range of wavelengths, and neutral gases near stars are subject to photoionization, even at the distance of 1 AU (cf. at Earth's ionosphere).

Collision with high-energy particles is another important ionization source. Galactic cosmic ray represents a homogeneous background of particles in interstellar and interplanetary space, mostly consisting of energetic protons reaching 10^{20} eV. When galactic cosmic ray hits a neutral atom, electrons are kicked out of the atom. Electron shower or precipitation from the Earth's magnetosphere also maintains ionization in the ionosphere, and causes aurora in the polar region.

Charge exchange is another process of ionization. Electrons can be exchanged from neutral atoms to ions when they meet, while the momenta of these species are maintained. This process creates energetic neutral atoms (which were initially energetic ions) and provides a very useful diagnostic and visualization tool for studying planetary magnetospheres.

In pulsar magnetospheres, the magnetic field originating from the neutron stars is so strong and, furthermore, the field is moving and co-rotating with the neutron stars so fast that pairs of electrons and positrons are created from the moving field. This creates a special kind of plasma, as there is no mass difference between two different particle species (cf. the proton mass is much larger than the electron mass, by factor 1836).

2.1.1.2 Fundamental Properties of Space Plasmas

Quasi-neutrality

In plasma, the electrostatic potential of individual ion is shielded by ambient, freely moving electrons. The potential curve does not fall as $1/r$ from the source (Coulomb potential), but falls faster as $e^{-r/\lambda_D}/r$ (Yukawa-type potential). Debye length λ_D characterizes the cutoff length scale of the potential and is a function of the electron temperature T_e and the number density n_e, $\lambda_D = \sqrt{\varepsilon_0 k_B T_e / n_e e^2}$, where ε_0 denotes the vacuum dielectric constant. Plasmas are quasi-neutral on spatial scales larger than Debye length (e.g., in the solar wind the Debye length is of order of 10 m), which justifies quasi-neutrality in plasma. Macroscopic properties such as the density, bulk velocity, and temperature often give a more convenient way of describing plasma dynamics. For the quasi-neutrality of plasma the energy density of electric field is found to be smaller than that of magnetic field (cf. Eq. 2.31).

Electrons are oscillating around ions due to the electrostatic attraction. This phenomenon gives the fundamental time scale, called the plasma oscillation, and its frequency is determined solely by the electron number density, $\omega_{pe} = \sqrt{n_e e^2 / m_e \varepsilon_0}$. In the solar wind plasma, frequency ω_{pe} is of order of 10 kHz.

High Conductivity

Due to mobility of electrons and ions, plasma is electrically conducting. In a highly conducting medium (gas or fluid) the magnetic field is said to be "frozen-in" to the plasma, that is the field moves together with the medium. On the other hand, for small-scale or high-frequency fluctuating fields the response of the ions and electrons is frequency-dependent, and conductivity on the fluctuating electric field is also frequency-dependent. For this reason there are a number of wave modes or normal modes in plasma, some of which are associated with the bulk motion, and others with the individual ion motion or electron motion.

Collisionless gas

The huge spatial scale and the low density in the interplanetary or interstellar space make plasmas collisionless; the mean-free-path of the particles is comparable to the system size or even larger than that. This fact raises questions with regard to the physical processes of shock waves and turbulence, as they need dissipation mechanisms to heat the plasma and to convert the energy into the thermal one. In ordinary gas dynamics, in contrast, dissipation is provided by binary collisions between particles.

The collisionless property of the space plasma also makes it possible to sustain thermally non-equilibrium states for a longer time, i.e., a non-Maxwellian velocity distribution does not evolve into Maxwellian but evolves in a different way. Examples of non-Maxwellian distributions are counter-streaming beams upstream of the shock wave (the foreshock region, Fig. 1.4) and anisotropic velocity distribution between parallel and perpendicular directions to the magnetic field (temperature anisotropy) downstream of the shock wave.

The concept of temperature can be generalized and applied to a collisionless gas even though it is not in thermal equilibrium: The kinetic temperature is a measure of the spread of the velocity distribution. Because of the collisionless property, a plasma can have different temperatures at the same time; The kinetic temperature (or the spread of the velocity distribution) can be defined and measured in various directions parallel and perpendicular to the magnetic field. Furthermore, electrons and ions can have different temperatures.

Nonlinearity

Plasmas exhibit various kinds of nonlinear phenomena. Even in the one-fluid picture of plasma, called magnetohydrodynamics, there are different sources of nonlinearity originating in the fluid motion (advection), the Lorenz force acting on the plasma, and the coupling between the flow velocity and the magnetic field. Nonlinear effects are, for example, wave steepening or wave-wave interactions that lead the plasma to evolve into turbulence. Under some circumstances, on the other hand, the equations of plasma dynamics can be reduced into a simpler integrable form. In such a case nonlinear effects compete against other effects such as dispersion (which means the broadening of wave packet) and the plasma dynamics leads to formation of large-amplitude, stable solitary wave.

2.1.2 Magnetohydrodynamics

Three Pictures of Plasmas

Plasmas are collections of a very large number of electrically charged particles, and plasma dynamics can be treated and described in different ways. Roughly speaking, there are three approaches: (1) single particle approach; (2) kinetic approach; and (3) fluid approach.

In the single particle approach, the motion of individual particle is considered. Electric field, magnetic field, and current are computed by counting individual particles. The conceptual structure of the method is relatively straightforward: solving equations of motions for each particle under Coulomb force, Lorentz force, and other forces; and then determining electric and magnetic fields using Maxwell equations using the charge density and the electric current. Charged particles exhibit various dynamics: gyration around the magnetic field; drift motions of the guiding center (of gyrating particles) under the electric field, gravity, non-uniform magnetic field, and curved magnetic field lines.

In the kinetic approach, evolution of phase space density (spanned by spatial coordinate and velocity coordinate) or velocity distribution function is studied using Boltzmann-Vlasov equation, which comes from Liouville's theorem stating the conservation law of phase space density. Plasma dynamics is treated in a statistical fashion, and this approach can resolve wave-particle interactions (instabilities, wave excitation, wave damping).

In the fluid approach, plasma is treated as an electrically conducting, magnetized, fluid characterized by the macroscopic quantities such as density, bulk velocity, and temperature. One may consider plasmas as consisting of multi-fluids (electron fluid and ion fluid; or fluids for different particle species), or more simply as one fluid. In the latter case the treatment is called magnetohydrodynamics (MHD).

Fluid Picture

Macroscopic variables in the fluid picture are closely related to the velocity distribution of particles. Number density n or mass density ρ is given as the integrated velocity distribution over the velocity coordinate, i.e., total volume of the velocity distribution. The bulk velocity or the flow velocity, $\mathbf{u}$, is given as the first-order velocity moment of the distribution, i.e., the center of the distribution or the average of individual particle velocities. The thermal velocity and pressure are given as the second-order moment of the velocity distribution, i.e., the spread of the distribution.

MHD essentially consists of two equation sets: flow dynamics for a conducting medium and electromagnetism. These equations can be written as follows.

$$\frac{\partial \rho}{\partial t} + \nabla \cdot (\rho \mathbf{u}) = 0 \tag{2.2}$$

$$\frac{\partial (\rho \mathbf{u})}{\partial t} + \nabla \cdot (\rho \mathbf{u}\mathbf{u}) = -\nabla \cdot \mathsf{P} + \mathbf{j} \times \mathbf{B} + \rho \nu \nabla^2 \mathbf{u} \tag{2.3}$$

$$\nabla \times \mathbf{B} = \mu_0 \mathbf{j} \tag{2.4}$$

$$\nabla \times \mathbf{E} = -\frac{\partial \mathbf{B}}{\partial t} \tag{2.5}$$

$$\nabla \cdot \mathbf{B} = 0, \tag{2.6}$$

where ν denotes the viscosity. P is the three-by-three pressure tensor, and each diagonal component represents the thermal pressure in different directions: for example, one parallel direction to the magnetic field and two perpendicular directions. One of the Maxwell equations is trivial under quasi-neutrality, $\nabla \cdot \mathbf{E} \simeq 0$. In MHD equations, the time derivative of the electric field is regarded to be so slow that the validity of the one-fluid approximation is not violated.

The above set of equations is not complete but has a closure problem associated with the thermal pressure. It cannot be determined within the above set of equations, and needs to be evaluated by other means. One possible way is to use the energy balance equation. Another possible way is to use the equation of state, relating the pressure to the density. Or one might assume the isothermal condition under which the thermal pressure is constant.

The electric current density can be evaluated either from the spatial variation of the magnetic field (Ampère's law) or from generalized Ohm's law. In a plasma Ohm's law becomes considerably more complicated, and contain different contributions to the electric current or to the electric field. Generalized Ohm's law is given as:

$$\mathbf{E} + \mathbf{u} \times \mathbf{B} = \frac{1}{\sigma}\mathbf{j} + \frac{1}{ne}\mathbf{j} \times \mathbf{B} - \frac{1}{ne}\nabla \cdot \mathsf{P}_e + \frac{m_e}{ne^2}\frac{\partial \mathbf{j}}{\partial t}. \tag{2.7}$$

Here σ denotes the conductivity. The first term on the right-hand-side represents the resistive, Ohmic term; the second term the Hall term; the third term the anisotropic electron pressure; and the last term the contribution of electron inertia to the current flow.

Incompressible MHD Equations

In the incompressible case, $\nabla \cdot \mathbf{u} = 0$, the MHD equations have a more symmetric structure between the magnetic field and the flow velocity. After arranging terms, the equations are expressed as follows.

$$\left(\frac{\partial}{\partial t} - \nu \nabla^2\right) \mathbf{u} = -(\mathbf{u} \cdot \nabla)\mathbf{u} + (\mathbf{b} \cdot \nabla)\mathbf{b} - \nabla P_{\text{tot}} \qquad (2.8)$$

$$\left(\frac{\partial}{\partial t} - \eta \nabla^2\right) \mathbf{b} = -(\mathbf{u} \cdot \nabla)\mathbf{b} + (\mathbf{b} \cdot \nabla)\mathbf{u} \qquad (2.9)$$

$$\nabla \cdot \mathbf{u} = 0 \qquad (2.10)$$

$$\nabla \cdot \mathbf{b} = 0, \qquad (2.11)$$

where the magnetic field is expressed in units of velocity, $\mathbf{b} = \mathbf{B}/\sqrt{\mu_0 \rho}$. The mass density ρ is assumed to be constant and has been omitted here. The symbol η denotes the diffusivity, $\eta = 1/\mu_0 \sigma$. P_{tot} is the total scalar pressure which is the sum of thermal and magnetic pressure, $P_{\text{tot}} = p_{\text{th}} + B^2/2\mu_0$. The fluid momentum equation (Eq. 2.8 which is the Navier–Stokes equation for incompressible fluid including the effect of the Lorentz force acting on the fluid element) and the induction equation (Eq. 2.9) have a very similar structure to each other in incompressible MHD. The difference of the two equations are (1) the pressure gradient term for the total pressure (thermal and magnetic pressures) appears in the fluid equation but not in the induction equation; and (2) the nonlinear terms are self-coupling (velocity to velocity, and magnetic field to magnetic field) in the fluid equation, whereas the induction equation exhibits cross-coupled nonlinear terms (velocity to magnetic field). Derivation of the induction equation (Eq. 2.9) makes use of the expression of electric field which is a sum of the convective field and the Ohmic field (cf., Eq. 2.7)

$$\mathbf{E} = -\mathbf{u} \times \mathbf{B} + \frac{1}{\sigma}\mathbf{j}, \qquad (2.12)$$

and Ampère's law (Eq. 2.4). There are four nonlinear terms in incompressible MHD: two of them are self-coupling terms $(\mathbf{u} \cdot \nabla)\mathbf{u}$ (advection of the flow) and $(\mathbf{b} \cdot \nabla)\mathbf{b}$ (Lorentz force acting on the fluid) and the other two are cross-coupling terms $(\mathbf{u} \cdot \nabla)\mathbf{b}$ and $(\mathbf{b} \cdot \nabla)\mathbf{u}$ that stem from the convective electric field $\mathbf{E} = -\mathbf{u} \times \mathbf{B}$.

It is interesting to see that the magnetic field and the flow velocity are almost interchangable. Indeed, exchanging the two variables $\mathbf{u}$ and $\mathbf{b}$ such that

$$\mathbf{u} = \pm \mathbf{b} \qquad (2.13)$$

is one of special solutions (in the limit of constant total pressure, $v \to 0$, and $\eta \to 0$). This state is called the Alfvénic state and represents one of the minimum energy states in magnetohydrodynamics.

Magnetic Pressure and Tension

The Lorentz force acting on the fluid element $\mathbf{j} \times \mathbf{B}$ plays an important role in magnetohydrodynamics, introducing the concept of magnetic pressure and tension. On applying vector algebra to Ampère's law, the Lorentz force splits into two terms:

$$\mathbf{j} \times \mathbf{B} = -\nabla \left(\frac{B^2}{2\mu_0} \right) + \frac{1}{\mu_0} \nabla \cdot (\mathbf{BB}). \qquad (2.14)$$

The first term on the right-hand side suggests that the magnetic field has a pressure $p_{\text{mag}} = B^2/2\mu_0$ and its spatial gradient serves as a force. The magnetic pressure simply adds to the thermal pressure of the plasma. The second term is a consequence of the vector product of current and magnetic field and thus of the vector character of the magnetic field. It is the divergence of a magnetic stress tensor $\mathbf{BB}/\mu_0$. The curved magnetic field lines introduces a magnetic stress in the plasma, which contributes to tension and torsion in the conducting fluid.

The advection term, $\rho (\mathbf{u} \cdot \nabla)\mathbf{u}$, can also be expressed in the form of pressure gradient as $-\nabla \rho u^2/2$ when the flow is not helical, $\mathbf{u} \cdot \nabla \times \mathbf{u} = 0$. If we neglect the off-diagonal or stress terms in the pressure tensors, e.g., for cases where the plasma pressure is nearly isotropic and the magnetic field is approximately homogeneous, the pressure equilibrium can be expressed as

$$\nabla \left(\frac{\rho u^2}{2} + p_{\text{th}} + \frac{B^2}{2\mu_0} \right) = 0. \qquad (2.15)$$

This means that in an equilibrium, isotropic, and quasi-neutral plasma the total pressure is a constant. Under these conditions one can define a plasma parameter *beta* as the ratio of thermal to magnetic pressure:

$$\beta = \frac{p_{th}}{p_B} = \frac{p_{th}}{B^2/2\mu_0}. \qquad (2.16)$$

In anisotropic plasmas where the pressure has distinct parallel and perpendicular components the concept of plasma beta can be applied to these components separately. The parameter beta measures the relative importance of gas and magnetic field pressures. A plasma is called a low-beta plasma when $\beta \ll 1$ and a high-beta plasma for $\beta \sim 1$ or greater.

Frozen-in Magnetic Field

The induction equation shows that the magnetic field evolves in two different fashions: (1) the field is coupled to the plasma motion (or the flow velocity), referred

Fig. 2.2 Frozen-in magnetic
field lines

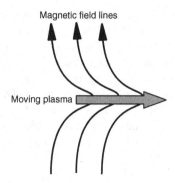

to as the frozen-in magnetic field, and (2) the field strength is subject to diffusion
and becomes weaker.

In order to study the transport of field lines and plasma more quantitatively, we
may use the simplified version of Ohm's law $\mathbf{j} = \sigma(\mathbf{E} + \mathbf{u} \times \mathbf{B})$, the induction
equation (Eq. 2.5) and the divergence-free condition of the field (Eq. 2.6). The time
evolution of the magnetic field is then expressed as

$$\frac{\partial \mathbf{B}}{\partial t} = \nabla \times (\mathbf{u} \times \mathbf{B}) + \eta \nabla^2 \mathbf{B}. \tag{2.17}$$

In the case where the diffusivity η is negligible the magnetic field is said to be
"frozen-in" into the plasma. The induction equation reduces into the form

$$\frac{\partial \mathbf{B}}{\partial t} = \nabla \times (\mathbf{u} \times \mathbf{B}). \tag{2.18}$$

This is identical with the equation for the coupling of the vorticity ($\nabla \times \mathbf{u}$) in
inviscid fluids, in which any vortex lines move together with the fluid. Equation
(2.18) implies also that the magnetic field lines are constrained to move with the
plasma. For example, if a patch of plasma populating in a bundle of field lines moves
perpendicular to the field, the magnetic field lines will be deformed in the manner
shown in Fig. 2.2.

It can also be shown that Eq. (2.18) implies that the total magnetic induction (or
the field lines) encircled by a closed loop remains unchanged even if each point on
this closed loop moves with a different local velocity. The concept of the frozen-
in magnetic field can actually be identified by the plasma glued to the field. The
equivalent expression is that the electric field is solely convective, determined by the
flow velocity and the magnetic field:

$$\mathbf{E} + \mathbf{u} \times \mathbf{B} = 0. \tag{2.19}$$

Equation (2.19) implies that in an infinitely conducting plasma there is no electric
field in the frame co-moving with the plasma. Electric field can only result from
the Lorentz transformation. Moreover, Eq. (2.19) contains another important point.

Since the cross-product between any velocity component parallel to the magnetic field and the field itself is zero, we can immediately see that any component of the electric field parallel to the magnetic field must vanish in an infinitely conducting plasma. An important consequence of the frozen-in magnetic field is that plasmas of different origins and attached to different magnetic field lines cannot be mixed under the frozen-in condition. Different plasmas may touch each other by forming a boundary or a current sheet but they are not mixed in the fluid picture.

In the case that the resistivity term dominates the evolution of magnetic field, Eq. (2.17) becomes a diffusion equation:

$$\frac{\partial \mathbf{B}}{\partial t} = \eta \nabla^2 \mathbf{B}. \tag{2.20}$$

Under the influence of a finite resistance in the plasma, the magnetic field tends to diffuse across the plasma and to smooth out any local inhomogeneities. The characteristic time of magnetic field diffusion is found by replacing the vector derivative by the inverse of the characteristic gradient scale of the magnetic field, L. The local solution of the diffusion equation is then:

$$B = B_0 e^{-t/\tau_d}, \tag{2.21}$$

where τ_d is the magnetic diffusion time given by

$$\tau_d = \frac{L^2}{\eta}. \tag{2.22}$$

Whenever $\eta \to 0$ or when the characteristic gradient length L is very large, the decay or diffusion time of the magnetic field becomes longer.

For an example consider the solar wind. Its density is of the order of $5\,\text{cm}^{-3}$ and its electron temperature is about $50\,\text{eV}$. The estimated magnetic diffusion time is scaled as $\tau_d \simeq 0.3L^2$ for collisions between electrons and ions, where τ_d is given in units of seconds and L in meters. The time that the solar wind needs to flow with its typical velocity of $500\,\text{km/s}$ across the Sun–Earth distance of $1.5 \times 10^{11}\,\text{m}$ is $\tau_{sw} = 3 \times 10^5\,\text{s}$ or $3.5\,\text{days}$. Thus, the magnetic field is permitted to diffuse across the solar wind over the short distance of only about $1.9\sqrt{\tau_{sw}} \sim 10^3\,\text{m}$ during the travel across the Sun–Earth distance. The magnetic field is practically frozen-in to the solar wind and the field lines are carried by the flow.

The dominance of either the frozen-in field or the diffusion process can be measured by the magnetic Reynolds number. Rewrite the induction equation (Eq. 2.17) using the characteristic time and length scales:

$$\frac{B}{\tau} = \frac{UB}{L} + \frac{\eta B}{L^2}. \tag{2.23}$$

In this equation B is the average magnetic field strength and U represents the average plasma velocity perpendicular to the field, while τ denotes the characteristic time of magnetic field variations, and L is again the characteristic length over which the field varies. The ratio of the first to the second term yields the magnetic Reynolds number:

$$R_m = \frac{LU}{\eta}. \tag{2.24}$$

This is useful in deciding whether the magnetic field evolution is dominated by the diffusion process ($R_m \ll 1$) or the frozen-in process ($R_m \gg 1$). For example, the magnetic Reynolds number in the solar wind is estimated to be about $R_m \sim 7 \times 10^{16}$, reflecting our previous argument of the negligible magnetic diffusion in the solar wind. Note that only the perpendicular velocity enters the frozen-in convective motion. Any flow parallel to the magnetic field has no consequences. When $R_m \ll 1$ or even $R_m \sim 1$, the diffusion becomes important and may dominate the evolution of the magnetic field. In most large-scale and dilute plasmas in the solar system and astrophysical systems the frozen-in magnetic field is a reasonable assumption in a wide range of time scales.

Four Energy Forms

Plasmas exhibit various kinds of dynamical behaviors and processes. It is useful to consider which kind of energy is involved in each process. There are four different forms of energy densities when treating the plasma as a fluid.

(1) The kinetic energy density of the flow:

$$\mathscr{E}_K = \frac{1}{2}\rho u^2 \tag{2.25}$$

(2) The thermal energy density or thermal pressure:

$$\mathscr{E}_{th} = p_{th} = nk_B T \tag{2.26}$$

where we applied the ideal gas law but other choices of equation of state (and therefore the plasma model) are also possible.

(3) The magnetic energy density:

$$\mathscr{E}_M = \frac{B^2}{2\mu_0} \tag{2.27}$$

(4) The electric energy density:

$$\mathscr{E}_E = \frac{\varepsilon_0 E^2}{2}. \tag{2.28}$$

It is worthwhile to note that the ideal gas law or the equation of state relating the pressure by the density and temperature closes the equation set of magnetohydrodynamics. The thermal pressure and temperature may be anisotropic between the directions parallel and perpendicular to the magnetic field. The kinetic and thermal energy densities represent the gas dynamical properties of the plasma.

Quasi-neutrality in the plasma suggests that the magnetic energy density is larger than that for the electric field. In fact, the Fourier transform of the induction equation yields the following expression:

$$\omega \delta \mathbf{B} = -\mathbf{k} \times \delta \mathbf{E}, \tag{2.29}$$

where ω and $\mathbf{k}$ denote the frequency and the wave vector of the wave or the field disturbance. The ratio of the electric to magnetic field amplitude is thus proportional to the propagation speed (or the phase speed) in the plasma,

$$\left| \frac{\delta E}{\delta B} \right| = \frac{\omega}{k} = v_{ph}. \tag{2.30}$$

Under the non-relativistic treatment, where $v_{ph} < c$ (the speed of light is expressed by c), the ratio of the electric to magnetic energy is much smaller than unity,

$$\frac{\mathscr{E}_E}{\mathscr{E}_M} = \frac{\varepsilon_0 E^2 / 2}{B^2 / 2\mu_0} = \frac{v_{ph}^2}{c^2} \ll 1. \tag{2.31}$$

Validity of Magnetohydrodynamics

Magnetodydrodynamics is the one-fluid treatment of plasma and there is no distinction between the different particle species of the plasma. This approximation requires that time scale of variation of the plasma and fields must be longer than the characteristic time scale of the constituting particles, in particular ions. The characteristic frequency, ω, of any change must be smaller than the ion cyclotron frequency for the magnetohydrodynamic picture to be valid. This argument applies to the length scale, too. The magnetohydrodynamic picture describes plasma dynamics on the length scales larger than the ion gyro-radius. Magnetohydrodynamics is therefore restricted to slow time variations and large spatial scales. At such low frequencies the displacement current can safely be neglected in the Maxwell equations.

2.1.3 Fundamental Plasma Processes

Dynamics of plasma is diverse. Gas dynamical motion is coupled with electromagnetic field and the energy can be stored in different forms: kinetic energy, thermal energy, magnetic and electric field energy. Examples of plasma dynamics are waves and instabilities, turbulence, shock waves, and magnetic reconnection. Some of these processes stem from the gas dynamical properties such as shock waves, and others from the coupling between the plasma and the electromagnetic field.

2.1.3.1 Waves and Instabilities

In contrast to incompressible fluids, in which the perturbation mode is solely non-propagating eddies, MHD supports several types of waves or linear modes even in the incompressible limit. Consider the simplest case of a homogeneous plasma described by a constant thermal pressure and a constant density in a magnetic field. For sufficiently small perturbations we can linearize the MHD equations by splitting the pressure, density, and the magnetic field into the large-scale, background part and the fluctuating field. Furthermore, the fluctuating fields are assumed to be a plane wave characterized by the angular frequency ω and the wave vector $\mathbf{k}$ (Fourier transform), reducing the differential operators into algebraic expressions. The linearized MHD equations support three types of oscillation modes: the Alfvén mode, the fast and slow magnetosonic modes. The three MHD modes are non-dispersive waves, and the phase and group velocities are independent of frequencies. This reflects the fact that the one-fluid picture of the plasma, MHD, does not contain any intrinsic scales in the time nor spatial domain.

Linear MHD Modes

The Alfvén mode is characterized by the dispersion relation:

$$\omega^2 = k_{\parallel}^2 v_A^2, \tag{2.32}$$

where $k_{\parallel}$ denotes the parallel component of the wave vector with respect to the background magnetic field, and v_A denotes the Alfvén speed defined as $v_A = B_0/\sqrt{\mu_0 \rho_0}$. In this mode the plasma motion is incompressible and transverse to the background magnetic field. So is the magnetic field perturbation. The oscillation is maintained by the magnetic tension and the inertia of the plasma. The oscillating field lines may be analogously interpreted as an elastic string with bending shape (Fig. 2.3). The Alfvén wave propagates preferentially parallel or anti-parallel to the background magnetic field direction. In particular, the group speed of the wave is strictly aligned with the background field. The mode also exhibits correlation between the velocity fluctuation and the magnetic field fluctuation for anti-parallel propagation, and anti-correlation for parallel propagation.

The compressive waves are referred to as the fast and slow magnetosonic modes. The fast magnetosonic mode is characterized by the dispersion relation

$$\omega^2 = \frac{1}{2}k^2 \left[v_A^2 + c_s^2 + \sqrt{\left(v_A^2 + c_s^2\right)^2 - 4v_A^2 c_s^2 \cos^2 \theta} \right], \tag{2.33}$$

where c_s and θ denote the sound speed and the propagation angle with respect to the background magnetic field. A useful expression for the sound speed is $c_s = \sqrt{\gamma p_0/\rho_0}$, where γ is the polytropic index, p_0 the mean thermal pressure, and ρ_0 the mean mass density. The fast mode is compressible and the phase velocity is in the range $v_A^2 \leq v_{ph}^2 \leq v_A^2 + c_s^2$, being fastest for propagation perpendicular to the background magnetic field. The fluctuating field pattern of the fast mode is displayed

Fig. 2.3 Magnetic field lines
for Alfvén and fast
magnetosonic mode waves

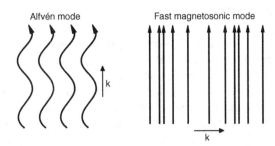

in Fig. 2.3. The magnetic field and the plasma are compressed by the wave motion, such that the restoring force is large and hence the frequency and the propagation speed are high. It is worthwhile to note that this mode exists even if the sound speed is zero.

The slow magnetosonic wave, or slow mode, is characterized by the dispersion relation

$$\omega^2 = \frac{1}{2}k^2\left[v_A^2 + c_s^2 - \sqrt{\left(v_A^2 + c_s^2\right)^2 - 4v_A^2 c_s^2 \cos^2\theta}\right].\qquad(2.34)$$

The phase velocity is in the range $0 \le v_{ph}^2 \le c_s^2$. The magnetic field oscillation and the pressure variation are anti-correlated with each other such that the restoring force acting on the medium is weaker than that for the fast mode. For this reason the frequency and the propagation speed are the lowest among the three MHD wave modes.

Non-MHD waves

On smaller scales or at higher frequencies the one-fluid treatment of plasma is no more valid. Electrons and ions react differently to the wave electric field, and the dielectric property of the medium for fluctuating field becomes a complex function of frequency. Therefore, plasmas allow a large number of wave modes to exist at higher frequencies. Waves can be treated in different fashions at high frequencies: cold plasma approximation using single particle motion or the kinetic treatment using Boltzmann-Vlasov equation. In both treatments waves are dispersive, i.e., the dispersion relation is no longer linear such that the propagation speed is dependent on frequencies.

For example, in the cold plasma treatment, there are two wave modes with the propagation direction parallel or anti-parallel to the background magnetic field: R-mode and L-mode with for their right-handed and left-handed polarization sense temporal evolution of field rotation sense. Here, polarization means the temporal field rotation sense when looking in the direction of the background magnetic field. The R-mode can exist at frequencies up to the electron gyro-frequency at which the wave is subject to the cyclotron resonance with electrons. The L-mode, on the other hand, exists at frequencies up to the ion gyro-frequencies, which is much lower than that for electrons. The L-mode is also referred to as the ion cyclotron mode.

Instabilities

Plasmas also exhibit various kinds of instabilities such that the force acting on the disturbed medium does not restore disturbance back to the equilibrium state but helps the fluctuation even to grow exponentially. In that case, the solution of the linearized set of equations for plasma dynamics is characterized by frequencies given as a complex number. The imaginary part of the complex frequency is interpreted as either wave growth or damping, depending on the sign of the imaginary part. Solutions with growing waves are referred to as instabilities. There are two types of instabilities: macroscopic and microscopic. Macro-instability originates in an unstable gradient or inhomogeneity of plasma in the coordinate space. An example for this is the Kelvin-Helmholtz instability in a shear flow configuration, generating eddies in the velocity shear. Micro-instability, in contrast, originates in an unstable gradient in the velocity space or unstable shape of the velocity distribution. Examples are a beam configuration showing two peaks in the velocity distribution or temperature anisotropy showing different thermal spread between parallel and perpendicular directions to the background magnetic field.

Wave growth or damping implies that the energy is exchanged between the plasma and the electromagnetic field. Two mechanisms are particularly known: cyclotron resonance and Landau resonance. While the former represents electromagnetic wave-particle interaction through a circularly polarized electric field, the latter represents electrostatic wave-particle interaction through a longitudinal electric field (along the background magnetic field). Both mechanisms accelerate electrons and ions in the direction of electric field.

2.1.3.2 Collisionless Shock Waves

Shock waves in ordinary gas dynamics are formed when a supersonic flow encounters an obstacle or when the object moves in a gas at a supersonic speed. The former is a standing shock wave (called the bow shock) where the shock propagation speed cancels out with the incoming flow speed. Space and astrophysical plasmas also exhibit shock waves in various places: solar atmosphere, interplanetary shocks, planetary bow shocks, supernova explosion. The nearest plasma shock to us is Earth's bow shock standing in the solar wind at about 15–20 R_E in front of the Earth. Most shock waves in space plasma are associated with the properties of fast mode: the flow speed must exceed the fast magnetosonic speed for the shock wave to occur, and both the gas pressure and the magnetic pressure are enhanced across the shock wave.

One of the fundamental differences in shock waves between ordinary gas dynamics and space plasma is the dissipation mechanism. While the shock waves in ordinary gas dynamics are maintained by binary collisions for the dissipation mechanism,

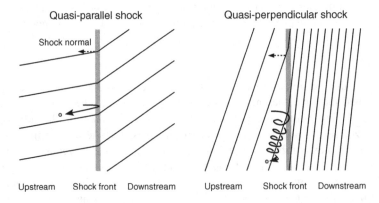

Fig. 2.4 Magnetic field lines across the quasi-parallel shock (*left*) and the quasi-perpendicular shock (*right*). At the quasi-parallel shock, charged particles are reflected at the shock front and stream backward along the field line. At the quasi-perpendicular shock, reflected particles cannot stream backward but gyrate around the magnetic field ahead of the shock front

those in space plasma cannot use binary collisions due to extremely large mean free paths. From in-situ observation of Earth's bow shock it was revealed that the plasma shock has a unique dissipation mechanism: particle reflection and acceleration, that is some particles in the incoming flow are accelerated and specularly reflected at the shock. Various mechanisms have been proposed for the reflection process: electric field at the shock wave repelling the particles; particle scattering in the shock-downstream region back to upstream; particle acceleration along the shock front. In any case, the phenomenon of particle reflection cannot be explained in the fluid picture of plasma, since not all particles are reflected but only a tiny portion of the incoming particle population.

Physical processes of the plasma shocks are further split into two distinct magnetic field geometries: quasi-parallel shock and quasi-perpendicular shock (Fig. 2.4). Here, the geometries refer to the angle between the upstream magnetic field direction and the shock normal direction.

At the quasi-parallel shock, the reflected particles escape into upstream along the magnetic field line and brake the incoming flow, warning the incoming flow about the existence of the shock wave. Hence the shock transition already starts with pre-thermalization in the upstream region, and this transition region is called the foreshock. Both ions and electrons are reflected and stream against the incoming flow. The foreshock plasma exhibits therefore a counter-streaming beam configuration: the incoming flow and the backstreaming beam. This forms an unstable velocity distribution, exciting waves. Across the shock the magnetic field lines are tilted more to the shock plane, which is the sense of field compression.

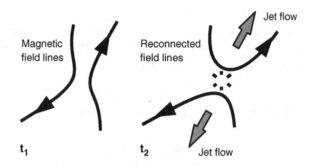

Fig. 2.5 Magnetic field lines before (t_1) and after reconnection t_2

At the quasi-perpendicular shock, the reflected particles gyrate around the upstream magnetic field direction and are swept by the flow back to the shock. The shock transition occurs therefore in a narrow layer consisting of the foot region ahead of the shock and the ramp on the scale of ion gyro-radius. Gyrating particles feel convective electric field (due to the flow) and are transported to the shock and further to the downstream region. For this reason the plasma is heated preferentially in the direction perpendicular to the magnetic field. The plasma downstream of the shock is characterized by temperature anisotropy (higher temperature in the perpendicular direction), which is subject to micro-instability, exciting the mirror mode. The magnetic field compression is more effective at the quasi-perpendicular shock.

2.1.3.3 Magnetic Reconnection and Dynamo

In plasmas energy conversion is possible between kinetic and magnetic energy. The magnetic reconnection is the process converting the energy from magnetic to kinetic, and the dynamo mechanism acts in the opposite sense from kinetic to magnetic. Magnetic reconnection occurs when the field configuration is anti-parallel across a thin current layer, and can be interpreted as the field lines cut and reconnected (Fig. 2.5). This process violates the frozen-in magnetic field and therefore requires local magnetic diffusion. The reconnected field lines have strong curvature and relaxes quickly by accelerating the plasma in two opposite directions as jet flow within a short time. A large amount of energy is released as the plasma jet flow (the conversion from the magnetic into the kinetic energy), and the magnetic reconnection is believed to be the very mechanism explaining explosion events in space plasma such as solar flares, coronal mass ejection, and geomagnetic substorms.

Dynamo is the process actively generating magnetic field in astrophysical bodies such as the Earth, the Sun, and other planets and stars. Without dynamo the magnetic field becomes weaker and decays due to the resistivity. Arguments for the need of the dynamo mechanism are: (1) The studies in palaeomagnetism revealed that Earth's magnetic field lasted longer than the decay time (the Earth outer core is not in the plasma state but in the liquid state, but it can be treated as an electrically conducting fluid and the use of magnetohydrodynamics is still valid) ; and (2) the solar magnetic field is oscillatory with the 11-year period. The essence in the dynamo mechanism is that the flow motion is preferable for twisting and stretching magnetic field lines such as helical flow motion due to buoyancy and Coriolis force, turbulence flow, and shear flow (or differential rotation).

2.2 Elements of Turbulence Theories

The set of equations of magnetohydrodynamics is highly nonlinear, yet the treatment of its turbulence can be found in fluid dynamics as a subset of magnetohydrodynamic equations (cf. similarity between the flow velocity and the magnetic field in incompressible MHD). The essence of fluid turbulence is reviewed as well as Kolmogorov's phenomenological model. Fluid turbulence is then compared with plasma turbulence.

2.2.1 Fluid Turbulence

In fluid dynamics two equations are primarily used: the continuity equation and the momentum balance equation representing conservation of mass and momentum, respectively. The continuity equation is the same as that in magnetohydrodynamics:

$$\frac{\partial \rho}{\partial t} + \nabla \cdot (\rho \mathbf{u}) = 0. \tag{2.35}$$

The momentum balance equation or Navier–Stokes equation (in a simplified form) is

$$\frac{\partial \mathbf{u}}{\partial t} + (\mathbf{u} \cdot \nabla)\mathbf{u} = -\nabla\left(\frac{p}{\rho}\right) + \nu\nabla^2\mathbf{u}, \tag{2.36}$$

where we assumed that the fluid is incompressible for simplicity satisfying the solenoidal condition:

$$\nabla \cdot \mathbf{u} = 0. \tag{2.37}$$

In other words, the mass density ρ is assumed to be constant. For an incompressible fluid the pressure p does not play an active role in dynamics, but it is *passive scalar* determined by the Poisson equation by taking divergence of Eq. (2.36). Navier–Stokes equation is nonlinear such that the flow velocity $\mathbf{u}$ is coupled to itself in the advection term. Although the Navier–Stokes equation appears to be relatively simple (there is only one nonlinear term in the incompressible fluid), its general solution is not known, and moreover, the existence and uniqueness of solution is not known, either. It is important to note that turbulence is not a property of fluid material but a property of flow whenever the fluid is set into motion and the advection term dominates in the momentum balance.

The convective derivative (time-dependent and advection term) requires that the Navier–Stokes equation or fluid equation be constructed in any frame of inertia as far as the frame of reference is transformed by Galilean transformation; its form is invariant under Galilean coordinate system transformation. Advection term is the driver of turbulent motion in the flow, and turbulence is characterized by the concepts of nonlinearity, unpredictability, and energy cascade and dissipation.

Nonliearity

The advection term may be interpreted as wave steepening. This can be seen by setting the flow velocity field as a plane wave, $u \propto \sin kx$. The advection term then yields $\mathbf{u} \cdot \nabla \mathbf{u} \sim k \sin(2kx)/k$, implying that the wave number is doubled from k into $2k$. An eddy with the vorticity $\nabla \times \mathbf{u}$ becomes enhanced by this term, thus the effect of advection can also be interpreted as vortex stretching.

The ratio of the magnitude of the advection to the dissipation term is called the Reynolds number Re,

$$\mathrm{Re} = \frac{|(\mathbf{u} \cdot \nabla)\mathbf{u}|}{|v\nabla^2 \mathbf{u}|} \sim \frac{U^2/L}{vU/L^2} = \frac{UL}{v}, \tag{2.38}$$

where again U and L denote the characteristic speed and length of flow and we use replacement $\nabla \to 1/L$.

Transition from laminar (flow at a low Reynolds number) to turbulence (flow at a high Reynolds number) is illustrated in Fig. 2.6. Streamlines are smooth at low Reynolds numbers and become increasingly complicated with eddies on various scales at high Reynolds numbers.

Typical values of the kinematic viscosity of air is about $v = 0.1 \, \mathrm{cm}^2 \, \mathrm{s}^{-1}$. If we walk in the air on the ground, the Reynolds number is estimated to be of order $\mathrm{Re} \sim 10^4 (L = 10 \, \mathrm{cm}, U = 100 \, \mathrm{cm} \, \mathrm{s}^{-1})$. For a motor vehicle the Reynolds number is about $\mathrm{Re} \sim 10^6 (L = 10^2 \, \mathrm{cm}, U = 10^3 \, \mathrm{cm} \, \mathrm{s}^{-1})$. In geophysical and astrophysical systems both the length and the flow speed are large numbers, and the Reynolds number is also very large. In the outer core of the Earth, the solar convection zone, and galaxy of the Reynolds number is of order of 10^8, 10^{10}, and even 10^{11}, respectively.

Unpredictability

Turbulent flow is irregular and prediction of flow velocity is very difficult. It is chaos of fluid (chaos in the sense of chaotic behavior in a dynamical system). A slight difference at the initial time ends up with a totally different state. It is worthwhile to note that the momentum balance equation or Navier–Stokes equation is deterministic, and since the equation is given, it should be in principle possible to determine the future motion of the flow as we know what forces are acting on the fluid element and how string they are. However, it is virtually impossible to solve the equation exactly and to predict the flow velocity in future because an error or an uncertainty grows rapidly.

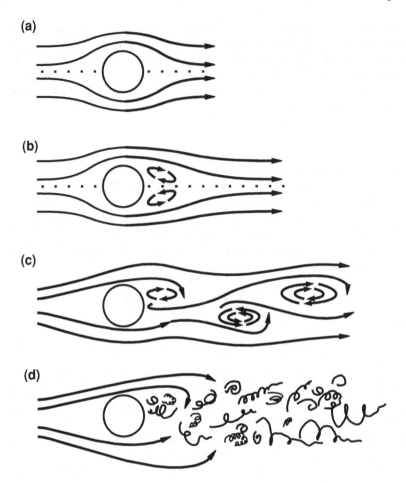

Fig. 2.6 Flow pattern around a cylinder at various Reynolds numbers: **a** 0.1; **b** 10; **c** 100; **d** 10000 (after Davidson, 2004, Turbulece: an introduction for scientists and engineers, fig. 1.4 p.8, by permission by Oxford University Press)

Energy cascade and dissipation

Nonlinear effects make it possible for different waves to interact with each other and to excite another wave mode. Wave steepening due to the advection term discussed above can be generalized such that the mode with the wave number k and the other mode with $2k$ are coupled to each other, and the wave–wave interaction proceeds successively in a cascade fashion. Turbulence is therefore characterized by the concept of energy cascade introduced in Chap. 1, in which large-scale eddies split into small-scale eddies, and further into eddies on even smaller scales. The energy flows from larger scales to smaller scales, and for this reason turbulent field does not show any characteristic size of eddies; different scales exist at once. For eddies on

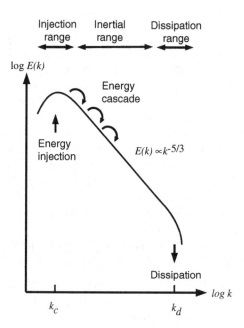

Fig. 2.7 Energy spectrum according to Kolmogorov's scaling law

larger scales the process of energy cascade serves as a sink, whereas for eddies on smaller scales it serves as a source. The concept of energy cascade can be formulated as a power-law energy spectrum (Fig. 2.7).

However, energy cascade cannot occur on arbitrary small scales. The reason for the existence of small-scale limit can be seen in the definition of the Reynolds number: on smaller scales the magnitude of the dissipation term becomes increasingly important. At sufficiently small wavelengths both terms (advection and dissipation) become of the same order in magnitude and the dissipation process will dominate and the kinetic energy of flow is converted to thermal energy. From the gas-dynamical point of view, this occurs on the scale of particle collisions, i.e., the mean free path. It is the dissipation process that heats the fluid on the smallest scale. In the stationary state of turbulence, energy is injected into the flow system on a large scale, which is transported to smaller scales by successive eddy splitting until the scale is reached where the curvature of the streamlines is so sharp that it is smoothed and the kinetic energy is converted into the thermal energy. If no energy is put into the flow system, energy cascade decays monotonously, and the flow becomes heated and smooth (freely decaying turbulence).

2.2.1.1 Phenomenological Model of Kolmogorov

The energy spectrum of turbulence was derived by Kolmogorov in 1941 [3, 4, 5]. His phenomenological derivation is based on the picture that physical process of turbulence is divided into three different ranges or spatial scales: injection range, inertial range, and dissipation range. They are divided from one spatial scale to another, in other words the separation is made in the wave number domain. Figure 2.7 displays the energy spectrum of turbulence in the wave number domain. On the largest scales or at lower wave numbers around k_c the energy is put or injected into the system by shear flow, inhomogeneity, instabilities, or external force. This range is called the injection range, and contains most of the fluctuation energy. In contrast, on the smallest scales or at higher wave numbers around k_d the dissipation process due to the finite viscosity in the fluid dominates and the fluctuation energy (or kinetic energy of flow) is converted into thermal energy. This range is called the dissipation range. The spectral energy drops sharply at the wave number for the dissipation scale (of order of the mean free path). Kolmogorov found that these two ranges are separated in a turbulence flow and connected by an intermediate range called the inertial range.

Kolmogorov found an analytical expression of the inertial-range spectrum. Its significance can be summarized as follows.

(1) The inertial range represents energy cascade or eddies splitting, transporting energy into smaller scales. It is the *inertia* of the turbulent flow in the wave number domain.
(2) The existence of the inertial range is universal in the sense that it is the property of the flow and independent from the type of fluid. It appears whenever the injection range and the dissipation range are well separated.
(3) The spectral curve must be power-law characterized by the exponent α, i.e., the energy spectrum is given as a function of the wave number, $E(k) \propto k^{-\alpha}$. This reflects the fact that the fluid equation has, in the limit of vanishing viscosity, symmetry with respect to the scaling transformation such that the length L and time T are scaled by λL and $\lambda^{1-h} T$, respectively, using the scaling factor λ and its exponent h. Scaling symmetry implies that turbulence does not exhibit any characteristic scale in fluctuation.
(4) The power-law exponent of the inertial-range spectrum is uniquely determined for isotropic, hydrodynamic turbulence, $\alpha = -5/3$.

Kolmogorov's phenomenological model is based on several assumptions: Fluctuation energy is transported by eddy splitting successively onto smaller scales; Equilibrium or balance holds in the energy transfer rate from the injection range (on the largest scale) down to the dissipation range (the smallest scale). The model uses two concepts in deriving the inertial-range spectrum: the energy transfer time and the energy transfer rate.

In the first step, the energy transfer time (or the eddy turnover time) is defined as

$$\tau_{ed} = l/v, \tag{2.39}$$

which implies that an eddy characterized by the size l and the flow speed v breaks up into small-scale eddies after time l/v, i.e., after the fluid element along the eddy streamline finishes just one circulation. We divide the scales discretely and write the length and the velocity scales as $l_1, l_2, \ldots, l_n$ and $v_1, v_2, \ldots, v_n$. The energy transfer time at n-th generation of cascade is then:

$$\tau_n = \frac{l_n}{v_n}. \tag{2.40}$$

In the second step, the energy transfer rate is estimated. The energy (per unit mass) at n-th generation is given by

$$E_n = \frac{1}{2} v_n^2. \tag{2.41}$$

The energy transfer rate defined as the spectral energy flux at n-th generation of energy cascade is

$$\varepsilon_n = \frac{dE_n}{dt} \sim \frac{E_n}{\tau_n}. \tag{2.42}$$

Here we replaced the differentiation by division for the argument of scaling-law. It is modeled such that all the energy stored at the length l_n and the velocity v_n is transported to the next generation or to a smaller scale by eddy breakup within the turnover time τ_n. Note that energy spectrum represents the energy density in the wave number domain. Since E_n has the units of squared velocity, the energy spectrum $E(k_n)$ must have the units of squared velocity per wave number interval Δk_n,

$$E_n = E(k_n) \Delta k_n. \tag{2.43}$$

In the third step, we use the equipartition of the energy transfer rate for turbulence in the stationary state. It is constant throughout the injection range, the inertial range, and the dissipation range:

$$\varepsilon_{\text{inj}} = \varepsilon_1 = \varepsilon_2 = \cdots = \varepsilon_n = \cdots = \varepsilon_{\text{dis}} = \text{const.} \tag{2.44}$$

Here ε_{inj} denotes the energy input in the injection range, and ε_{dis} the energy dissipated by viscosity on the smallest scale.

In the fourth step, we derive the scaling-law for velocity and energy spectrum. The energy transfer rate can be expressed by the length l_n and the velocity v_n when combined with the energy transfer time τ_n, and furthermore it is assumed to be constant: $\varepsilon_n = v_n^3/l_n = \text{const.}$ This suggests that the velocity is scaled using the length as $v = (\varepsilon l)^{1/3}$. The inertial-range energy spectrum can then be derived by comparing two different expressions for fluctuation energy. One is the energy scaled to the wave number:

$$E_n = \frac{1}{2} v_n^2 = \varepsilon^{2/3} k_n^{-2/3}. \tag{2.45}$$

The other is the expression using the spectral energy density:

$$E_n = E(k_n)\Delta k_n = C_K E(k_n)k_n, \tag{2.46}$$

where C_K denotes a coefficient that cannot be determined in the argument of scaling-law alone. The former equation uses the scaling law of velocity and the latter is a definition of the energy spectrum. Here we used the relation $\Delta k \propto k$, that is the wave number interval is equidistant on the logarithmic scale, $\log(\Delta k) = $ const. Hence we obtain the energy spectrum, omitting the subscript n, as

$$E(k_n) = C_K \varepsilon^{2/3} k^{-5/3}. \tag{2.47}$$

This is the inertial-range spectrum. Various experiments and observations including wind tunnel experiments, water channel experiments, and atmosphere and ocean observations confirm that Kolmogorov's energy spectrum is valid in those measurements, and therefore realization of Kolmogorov's scaling or energy spectrum is one of the central topics in studying turbulence. The coefficient C_K is called the Kolmogorov constant. Various experiments of fluid turbulence suggests that $C_K \simeq 1.6$.

Kolmogorov's inertial-range spectrum can also be derived from dimensional analysis. Energy (per unit mass) is the squared amplitude and its spectrum has the dimension energy divided by wave number, $[E] = L^3 T^{-2}$. The wave number has the dimension $[k] = L^{-1}$. The energy transfer rate has the dimension $[\varepsilon] = [v^2/\tau] = L^2 T^{-3}$. If we use *Ansatz* with a dimensionless coefficient C as

$$E(k) = C\varepsilon^\alpha k^\beta, \tag{2.48}$$

we obtain $\alpha = 2/3$ and $\beta = -5/3$ and reproduce the energy spectrum of Kolmogorov.

While Kolmogorov's phenomenology successfully describes the energy spectrum of fluid turbulence, it should in principle be possible to derive this spectrum by solving the Navier–Stokes equation directly. Such a theoretical approach leads us to the so-called closure problem of turbulence. The advection term in the Navier–Stokes equation has the quadratic form with respect to the flow velocity $\mathbf{u} \cdot \nabla \mathbf{u}$ if explicitly written, where $\mathbf{u}$ is the flow velocity, and the statistical treatment of the Navier–Stokes equation yields the following problem. Computation of the second order moments such as energy ($\propto u^2$) and helicity ($\propto \mathbf{u} \cdot \nabla \times \mathbf{u}$) requires the knowledge of the third order moments because the advection term gives the third order moments (i.e., transport of the kinetic energy) when the equation is multiplied by the velocity $\mathbf{u}$, and solving the equation for the third order moments now requires the knowledge of the fourth order moments, again because of the advection term, and the problem is repeated at ever higher order moments. There appears always a higher order term to determine moments of any order. Therefore the statistical treatment of the fluid dynamics equation cannot be not closed without using any assumptions or approximations. It is worthwhile to note that a similar closure problem happens when one derives the set of fluid equations (continuity and momentum equations at the lowest orders) from the Liouville equation by means of the velocity moment. The reason is due to the transport term in the Liouville equation, and the fluid equation system can be closed when introducing, for example, the equation of state to relate the density (the zero-th order velocity moment) by the pressure (the second-order moment). The closure problem of turbulence is about the statistics of the random fluctuating bulk velocity, and it should not be confused with the closure problem of the Liouville equation (which is about the statistics of individual particle motions). One possible remedy about the turbulence closure problem is to assume some analytical form for expressing the second order moment and close the equation at the level of the second order, for example using the concept of mixing length. This family of theories is called the one-point closure theory. Using the mean-field decomposition method separating the velocity field into a mean field part and a fluctuating field part, the turbulence effect is expressed as the Reynolds stress tensor, a three-by-three correlation tensor of the fluctuating velocity field, and the essential task in the turbulence closure problem is to find a reasonable expression of this correlation tensor. The mean-field picture of turbulence also suggests that velocity shear or gradient on a large scale serves as the driver of turbulence, injecting the kinetic energy into the system, and then the injected energy is transferred from a large scale to a small scale by the action of the nonlinear effects of the flow.

Another family of the statistical theories of turbulence involves two-point closure of the correlation tensor, in which the moments of the Navier–Stokes equations are extended to higher orders until the fourth order moments appear, and then one uses a cumulant expansion so that the fourth order moments are replaced by products of the second order moments by assuming that the fluctuating field follows almost Gaussian distribution of probability (quasi-normal approximation). By using two more additional assumptions (eddy damping and Markovian process), it is possible to integrate the Navier–Stokes equation in the statistical sense and one obtains time evolution of energy and helicity spectra. Such a treatment with the three assumptions or approximations is referred to as the EDQNM theory, abbreviation of Eddy-Damped Quasi-Normal Markovian [7, 8, 9]. Further refinement in the two-point closure theory was initiated by Kraichnan, called Direct Interaction Approximation (DIA). In this method the higher order moments are evaluated using the method of renormalized Green functions [10]. The DIA method successfully reproduces Kolmogorov's energy spectrum as well as the Kolmogorov constant without introducing any arbitrary or adjustable parameters [11]. Although the DIA method demands a lot of calculation, it is regarded as the most refined theoretical approach in turbulence studies because it does not introduce any arbitrary or free parameter. Detailed calculations are also explained in Ref. [12] and [13].

The two-point closure theories are based on the assumption that statistics of fluctuating fields (i.e., probability distribution function) almost exhibits the Gaussian distribution. This assumption is useful because higher order moments are expressed by products of second order moments. In other words, second order moments (energy and helicity) determine dynamics and structure of turbulence. On the other hand, deviation from the Gaussian distribution is also an important feature in turbulence because it is the sign of non-vanishing higher order moments (or strictly speaking, higher order cumulants) and also the sign of wave–wave (or eddy-eddy) interactions. If the probability distribution function strictly followed the Gaussian distribution then fluctuations are composed of completely random and incoherent waves in the sense that the phases of the Fourier modes are randomly distributed and turbulence cannot continue energy cascade because the third order moment which is responsible for wave–wave interactions vanishes. Deviation from the Gaussian distribution can be recognized in real fluctuating fields as sparse, spiky signals. The deviation from the Gaussian distribution may be scale dependent such that one needs to establish a correction method in the energy cascade process to account for higher order structures or fine structures. Such spiky signals or localized fine structures are referred to as the intermittency, and it may be regarded as higher order correction of the statistical theory of turbulence. There are various kinds of intermittency models. Some deal with a log-normal distribution and others deal with a log-Poisson distribution.

Fluid turbulence is ultimately a heating process that occurs after energy cascade of eddies over many generations and decades of spatial scales. From the energy conservation viewpoint , the flow kinetic energy is transported from a large scale to a smaller scale by eddy splitting and then finally the kinetic energy is converted to the thermal energy in the dissipation range. Turbulence is the property of the flow and any fluid is subject to turbulence once it is set into motion under a high Reynolds number by some external force or inhomogeneities. The concept of viscosity, in contrast, is one of the properties of the fluid and is dependent on the molecular or other particle structure in the fluid. While gases in interplanetary and interstellar space can be treated as an application of fluid mechanics and it is natural to ask if these gases also exhibit turbulence behavior, one should keep in mind that these gases are ionized and electrically conducting, which gives gas dynamics more degrees of freedom and makes turbulence behavior more complex.

2.2.2 Plasma Turbulence

While fluid turbulence can be regarded as the energy conversion problem from kinetic to thermal energy, plasma turbulence is a more complex problem because four energy forms are involved: kinetic, thermal, electric, and magnetic energy. There are physical processes in plasma that convert energy from one type into another. For example, magnetic reconnection converts energy from magnetic to kinetic; wave damping converts energy from electric and magnetic into thermal energy. Energy cascade was solely carried by eddy splitting in ordinary fluid turbulence, since eddies are the only possible oscillatory motion in incompressible fluid mechanics, but in plasmas more complex wave–wave interaction and scattering processes may be the carrier of the energy cascade, in particular, Alfvén waves can exist for large amplitudes and therefore scattering of Alfvén waves are a good candidate of energy cascade in plasma turbulence. Dissipation is also a problem in plasma turbulence, since there is no particle collision. Instead of binary collisions between particles, wave damping or wave-particle interactions may work as the dissipation mechanism.

There are a variety of theoretical approaches to MHD turbulence. Some are phenomenological models and some follow the procedure of closure theories. One of the phenomenological models of MHD turbulence is based on the assumption of the

Alfvén wave scattering as the primary energy cascade process [14, 15] and predicts that the energy spectrum is expressed as the power-law to the wave number with the index $-3/2$. Other phenomenological models use energy cascade as competition between eddy splitting and Alfvén wave scattering [16, 17]. The energy spectrum in MHD turbulence depends on the Alfvén speed, the characteristic wave propagation speed in plasma. In contrast to the eddy splitting scenario of fluid turbulence, self-similarity of energy cascade in principle fails in MHD turbulence because of the Alfvén speed determined by the large-scale quantities (magnetic field strength and mass density) is involved.

Closure theories such as the EDQNM approximation can also be applied to the MHD turbulence problem. The EDQNM model suggests that five spectra are essential to characterize MHD turbulence: kinetic and magnetic energies; kinetic, magnetic, and cross helicities. It is possible, for example, to establish turbulence with a conserved magnetic helicity and vanishing cross helicity (called the helical MHD turbulence). It is also possible to establish turbulence with conserved cross helicity and vanishing magnetic helicity (called the dynamic alignment). MHD turbulence exhibits a higher degree of freedom than fluid turbulence and therefore makes the problem challenging. To date, no direct evidence is obtained in space plasma justifying the use of the phenomenological model or the closure theories of MHD turbulence. There are various reasons for this: theories are developed on several assumptions, and in particular, the assumption of isotropic turbulence is questionable, because the existence of a large-scale or mean magnetic field introduces a special direction in plasma; previous in-situ observations using spacecraft have been performed in the one-dimensional, temporal domain because of single point measurements in space and therefore observation could not distinguish between temporal and spatial variations.

References

1. Dodelson, S.: Modern Cosmology. Academic Press/Elsevier, San Diego (2003)
2. Chen, F. F.: Introduction to Plasma Physics and Controlled Fusion. Plenum Pub, New York (1984)
3. Kolmogorov, A.N.: The local structure of turbulence in incompressible viscous fluid for very large Reynolds number, Dokl. Akad. Nauk. SSSR 30, 299–303 (1941a) Reprinted in Proc. Roy. Soc. A 434, 9–13 (1991)
4. Kolmogorov, A.N.: Dissipation of energy in locally isotropic turbulence, Dok. Akad. Nauk. SSSR 32, 19–21 (1941b) Reprinted in Proc. Roy. Soc. A 434, 15–17 (1991)
5. Frisch, U.: Turbulence: The Legacy of A. N. Kolmogorov. Cambridge University Press, Cambridge (1995)
6. Davidson, P.A.: Turbulence-An Introduction for Scientists and Engineers. Oxford University Press, USA (2004)
7. Orszag, S.A.: Analytical theories of turbulence. J. Fluid Mech. 41:363–386 (1970)
8. Monin, A.S., Yaglom, A.M.: Statistical Fluid Mechanics: Mechanics of Turbulence. vol. 2, MIT Press, Cambridge (1975)
9. Lesieur, M.: Turbulence in Fluids, 3rd edn. Kluwer Academic Publishers, Dordrecht (1997)

10. Kraichnan, R.H.: The structure of isotropic turbulence at very high Reynolds numbers. J. Fluid Mech. **5**, 497–543 (1959)
11. Kraichnan, R.H.: Lagrangian-history closure approximation for turbulence. Phys. Fluids **8**, 575–598 (1965)
12. Leslie, D.C.: Developments in the Theory of Turbulence. Clarendon Press, Oxford (1973)
13. McComb, W.D.: The Physics of Fluid Turbulence. Clarendon Press, Oxford (1990)
14. Iroshnikov, P.S.: Turbulence of a conducting fluid in a strong magnetic field. Sov. Astron. **7**, 566–571 (1964)
15. Kraichnan, R.H.: Inertial range spectrum in hydromagnetic turbulence. Phys. Fluids **8**, 1385–1387 (1965)
16. Shridhar, S., Goldreich, P.: Toward a theory of interstellar turbulence, I. Weak Alfvénic turbulence. Astrophys. J. **432**, 612–621 (1994)
17. Goldreich, P., Shridhar, S.: Toward a theory of interstellar turbulence, II. Strong Alfvénic turbulence. Astrophys. J. **438**, 763–775 (1995)

Chapter 3
Multi-Spacecraft Measurements

Abstract Multi-spacecraft observation is a powerful method in turbulence research in space plasma. This chapter introduces analysis methods for Cluster data, which is a four-spacecraft mission observing waves and turbulence in near-Earth space plasma. Using multi-point measurements it is possible to distinguish between spatial structures and temporal variations in the data, and also to analyze the fluctuations in the rest frame of the flow, correcting for the Doppler shift. The principle of analysis techniques and applications such as energy spectra and dispersion relation are presented.

3.1 The Cluster Mission

3.1.1 Need for Multi-Point Measurements

Plasmas in the solar system such as in the planetary magnetospheres and the interplanetary space have been extensively studied in situ using spacecraft since 1960s. It is worthwhile to note that those missions are mostly single spacecraft missions. The data were obtained along the trajectory of the spacecraft and thus were of one-dimensional characteristics. Certainly single spacecraft data deliver a lot of useful information for studying space plasma dynamics: different plasma regimes and conditions (magnetosphere, solar wind), plasma boundaries (bow shock and magnetopause), magnetic field of other solar system bodies (planets and satellites), and so on. On the other hand, there still remain fundamental problems constrained by single-point measurements. In particular, distinguishing between temporal and spatial variation was not possible using single spacecraft data, and furthermore, due to the Doppler shift the analysis was limited to the spacecraft frame. For this reason, assumptions had to be used in the analysis, for example, Taylor's frozen-in flow hypothesis or dispersion relation for a specific wave mode to estimate wavelengths from frequency profile of the data. Separating the data into spatial and temporal variations and resolving spatial structures in three dimensions motivated physicists to propose the Cluster mission, the four-spacecraft mission.

Y. Narita, *Plasma Turbulence in the Solar System*, SpringerBriefs in Physics,
DOI: 10.1007/978-3-642-25667-7_3, © The Author(s) 2012

Single-point observations of fields such as plasma density, temperature, flow velocity, and magnetic and electric fields can conveniently be analyzed in the frequency domain, simply by Fourier transforming the time series data into the frequency domain. Power spectrum, for example, is a plot of field energy density (proportional to the squared amplitude) in the frequency domain, and it indicates frequencies at which the signals are mostly characterized. Yet, what does the frequency mean here? Does it really represent a purely temporal variation? Isn't there any spatial structure embedded in the time series data due to the background flow or the sensor motion?

These questions can clearly be formulated in the Doppler relation. Suppose we observe wave signal in the data. The measurement is performed at a fixed point in space but there is a flow in the medium (which is referred to as observer's frame or spacecraft frame). The background flow modulates the wave frequency by the Doppler shift, and the observed frequency is given as:

$$\omega' = \omega + \mathbf{k} \cdot \mathbf{V}, \tag{3.1}$$

where ω' and ω are the wave frequencies in observer's frame and in the flow rest frame (which is co-moving with the flow), respectively. The two reference frames are related to each other by Galilean transformation with a relative velocity $\mathbf{V}$ between the sensor and the flow. The wave vector $\mathbf{k}$ represents the spatial variation of the wave field. The wave vector is invariant under Galilean transformation, and only the frequency becomes modulated. The Doppler shift is modulation of the frequency and is given as a scalar product of the wave vector and the relative velocity.

Therefore the measurement in observer's frame include both intrinsic temporal variation (ω) and spatial variation convected or swept by the flow ($\mathbf{k} \cdot \mathbf{V}$) at once. Certainly, Doppler shift occurs when the sensor is moving in a rest medium, too. In order to correct the Doppler shift and to determine the intrinsic frequency ω, it is necessary to determine the wave vector and the flow velocity. In a general situation, the existence of dispersion relation and which kind of dispersion relation if any is not known, and the relation between frequencies and wave vectors may be arbitrary. In that case one has to determine the wave vector from the measurement. While the flow velocity and the spacecraft speed can be determined at single point, the wave vector can only be determined uniquely using multi-point measurements.

Taylor's Frozen-in Flow Hypothesis

A portion of information about the wave vector can still be obtained from single-point measurements under certain conditions. One method is to assume dispersion relation for observed waves. However, one of the goals of the data analyses is to determine the dispersion curve *experimentally*, and furthermore, it is not clear if the concept of dispersion relation is valid under strong turbulence or fully-developed turbulence. Therefore, the use of the assumption of dispersion relation cannot be justified in many situations. Another possible method for estimating spatial structure from single point data is to assume Taylor's frozen-in flow hypothesis. Suppose we measure a fluctuating field in a fast-streaming medium with the flow velocity $\mathbf{V}_f$. If the Doppler term $\mathbf{k} \cdot \mathbf{V}_f$ is much larger than the rest-frame frequency ω, the observer sees essentially the spatial variation characterized by the wave vector $\mathbf{k}$ as time series data with the temporal variation characterized by ω'. That is, if only intrinsic frequency could be neglected, we obtain from the Doppler relation the following equation:

$$\omega' \simeq \mathbf{k} \cdot \mathbf{V}_f. \tag{3.2}$$

This approximation is referred to as Taylor's frozen-in flow hypothesis [1]. The Doppler relation can be written in terms of velocities by dividing the relation by the magnitude of the wave vector,

$$\mathbf{v}'_{ph} = \mathbf{v}_{ph} + \mathbf{V}_f, \tag{3.3}$$

where $\mathbf{v}'_{ph}$ and $\mathbf{v}_{ph}$ are the phase velocities of the wave in the observer's frame and in the flow rest frame, respectively. Taylor's hypothesis essentially neglects the wave phase speed in the flow frame,

$$\mathbf{v}'_{ph} \simeq \mathbf{V}_f. \tag{3.4}$$

Taylor's hypothesis is a reasonable assumption in incompressible fluid turbulence because eddies do not propagate and represent zero-frequency mode. Taylor's hypothesis is also widely used in various studies of space plasma turbulence as it conveniently relabels the observed frequency into the wave number along the flow. But one has to bear in mind that waves in plasma can in general propagate, and may have the phase speeds comparable to the flow speed. Because of the limit of single-point measurements, spatial variations perpendicular to the flow cannot be obtained using Taylor's hypothesis. One possible way to estimate the three-dimensional spatial structure from single-point measurements is to incorporate further assumptions or approximations such as axisymmetry around the mean magnetic field or around the flow direction, or symmetry between parallel and anti-parallel directions to the magnetic field.

The Cluster Mission

The Cluster mission [2] is a four-spacecraft mission, and aims particularly in resolving spatial structures of complex space plasma dynamics in three dimension. Earlier spacecraft missions were limited to mostly one or at most two spacecraft and one could not distinguish between temporal and spatial variations, nor three-dimensional structure in space. The mission is capable of multi-point measurements with high time resolution and identical plasma and field instrumentation on board all the four satellites. After the first proposal in 1982 and beyond the tragic launch failure in 1996, Cluster was rebuilt and launched twice by the Soyuz-Fregat rockets in Baikonur Cosmodrome in 2000: two spacecrafts were first launched on 16 July, 2000, and later the other two spacecrafts on 9 August, 2000. Cluster has become

Fig. 3.1 Assembly of four Cluster spacecraft. Courtesy of European Space Agency

a revolutionary mission for studying space plasma such as near-Earth solar wind
and Earth's magnetosphere; performing formation flight in tetrahedron on various
sizes from over 10,000 km down to 100 km. Figure 3.1 displays the assembly of the
four Cluster spacecraft. Each spacecraft carries the same set of eleven instruments:
ASPOC (Active spacecraft potential control), CIS (Ion composition and energy spectrogram), EDI (Plasma drift velocity), FGM (DC and low-frequency magnetic field),
PEACE (Electron energy spectrogram), RAPID (High energy electrons and ions),
DWP (Wave processor), EFW (Electric field and waves), STAFF (Mid- to high-
frequency electric and magnetic field), WBD (Electric field and wave forms), and
WHISPER (Electron density and waves). The low-frequency magnetic field were
obtained by the FGM instrument (fluxgate magnetometer), and the data are primarily
used for studying low-frequency wave and turbulence phenomena in the solar wind
and in the regions upstream and downstream of Earth's bow shock, and in Earth's
magnetosphere.

Cluster forms a tetrahedral constellation with inter-spacecraft separation ranging
from 100 km to 20,000 km, depending on the mission phase, flying in a polar orbit
around the Earth with a perigee about 4 R_E and an apogee about 20 R_E. It is designed
to study small to large-scale structures and fluctuations in three dimensions in regions
such as the solar wind, the bow shock, the magnetopause, the polar cusps, the magnetotail, and the auroral zones. Figure 3.2 displays the highly elliptic, polar orbit of
Cluster in the spring season (February to March). Orbit period is 57 hours. The spring
season is very suitable for studying the solar wind and the Earth's bow shock, as the
spacecraft spend most of the time near the apogee (that is in the solar wind).

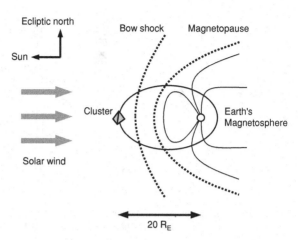

Fig. 3.2 Cluster orbit in the spring season (around February and March)

3.1.2 Spatial Aliasing

Aliasing is a general problem in the analysis of any measurement sampling at discrete points. Cluster performs sampling both in the temporal and spatial domains. While aliasing in the temporal domain can be suppressed by reducing high-resolution time series data using a suitable filter, sampling in the spatial domain causes a problem because the data are not processed by reduction of high-resolution spatial sampling using anti-aliasing filter. Sampling in the spatial domain results in a periodic pattern and furthermore distortion of energy spectra in the wave vector domain, and this effect is called spatial aliasing. The sampling problem is of particular importance for multi-spacecraft measurements in space, since one of the goals of the data analyses is to determine energy spectra in the wave vector domain. The theoretical background of aliasing problem after pioneering studies by Nyquist and Shannon, and its implications to spatially discrete measurements are summarized in [3, 4, 5].

It is true that multi-point measurements enable us to determine spatial structures but not on all spatial scales at once. For example, one cannot determine the spatial structure when its characteristic scale (e.g. wavelength) is smaller than the sensor separation distance. In that case the measurement makes under-sampling and the result is subject to the aliasing problem. Neubauer and Glassmeier presented the concept of spatial aliasing under multi-point measurements in space [6], and this issue was re-addressed using Cluster data [7, 8, 9, 10].

Figure 3.3 displays two examples of spatial aliasing from Cluster data in the solar wind. The left panel is the energy distribution in the wave vector domain. The plane is spanned by two reciprocal vectors of Cluster tetrahedron. The configuration of the sensor array can be represented as a set of reciprocal vectors in the wave vector domain, and the analysis should be performed within the first Brillouin zone or the Wigner-Seitz cell in the wave vector domain. The lattice structure in the energy distribution represents aliases. Within the first Brillouin zone, spanned by the half-length reciprocal vectors [11] there is no aliased spectrum. Beyond the

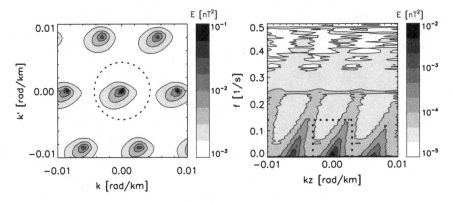

Fig. 3.3 Magnetic energy spectra showing aliases in the wave vector domain and in the frequency-wave vector domain, from Cluster data in the solar wind [10]

Brillouin zone in the wave vector domain the spectral signature does not reflect the true signal any more, or in other words, one cannot distinguish if the spectral signature is true or not beyond the Brillouin zone. Aliases of the principal distribution appear periodically in the wave vector domain, corresponding to the fact that one cannot resolve wavelengths that are smaller than the sensor separation. It is interesting to note that the configuration of sensor array may substantially influence the measurement of spatial structures when the tetrahedral shape is strongly irregular when the aliased spectra at larger wave numbers are extended such that they enter the first Brillouin zone from outside. One possible way to minimize the effect of spatial aliasing is to set a circle or a sphere in the wave vector domain that fits the Brillouin zone tangentially from inside (see dotted circle in the left panel).

Furthermore, when the spatially discrete measurements are performed in a flow (such as in the solar wind), spatial aliasing affects frequency profile of the energy spectra in observer's frame. The right panel in Fig. 3.3 displays such an example. The data are taken from the same time interval as in the left panel 'Cluster data in the solar wind), and processed by the multi-spacecraft data analysis (described later in this chapter) to derive the frequency-wave number profile of the energy spectrum along the flow. The spectrum exhibits a periodic structure in the wave number domain, which represents aliases. Due to the Doppler shift, the spectrum has a slope determined by the flow speed (note that the frequency-wave number slope represents the phase speed). At intermediate frequencies around 0.15 Hz the nearest-neighbor alias enters the first Brillouin zone from left, and the frequency profile of the spectrum is largely affected by spatial aliasing transmitted into frequency. This is because the Doppler shift affects the dispersion relation, and it transmits spatial aliasing into the time domain.

Aliasing is a general problem whenever discrete measurements sample continuous quantities. Periodic sampling of a continuous function results in a periodic structure in the Fourier domain. The spatial aliasing distorts the true spectrum when the aliases of the spectrum enters the Brillouin zone. Aliasing in the wave vector domain has two effects. One is weak aliasing, which yields anisotropy and asymmetry in the distribution. In this case the distortion represents primarily elongation of the distribution in the shortest lattice vector direction. The other effect is strong aliasing, in which the spectrum becomes dominated by the aliases. In the latter case, the distribution exhibits several peaks (original peak and aliased peak) in the case of strong aliasing.

3.2 Analysis Methods

3.2.1 Single-Point Methods

Measurements and analysis of waves and turbulence are often performed in the Fourier domain as fundamental wave properties such as frequencies and wave vectors spann the basis of the Fourier domain and other wave properties such as propagation directions and speeds, polarization, and amplitudes can be determined at various frequencies and wave vectors. The Fourier transform of magnetic field data from the temporal domain into the frequency domain is given as:

$$\tilde{\mathbf{B}}(f) = \int_0^T \mathbf{B}(t) \, e^{i2\pi f t} dt, \tag{3.5}$$

where $\mathbf{B}(t)$ is the magnetic field data at time t and $\tilde{\mathbf{B}}(f)$ is its Fourier transform. Note that the dimensions are different between the observed field and its Fourier transform. T denotes the time length of measurement. The cross spectral density (CSD) matrix is constructed from the Fourier transform as

$$\mathsf{R} = \frac{1}{T} \left\langle \tilde{\mathbf{B}}(f) \tilde{\mathbf{B}}^\dagger(f) \right\rangle. \tag{3.6}$$

Here the dimension of each matrix element is given in units of square amplitude per frequency (energy density in the frequency domain, nT^2/Hz.) The averaging (denoted by the symbol $\langle \cdots \rangle$) is made over many realizations, often replaced by averaging over many time intervals or sub-intervals, assuming that the average in the temporal domain is statistically representative (e.g., ergodic hypothesis). For the magnetic field data the CSD matrix is given as a three-by-three Hermitian matrix:

$$\mathsf{R} = \frac{1}{T} \begin{pmatrix} \langle \tilde{B}_x \tilde{B}_x^* \rangle & \langle \tilde{B}_y \tilde{B}_x^* \rangle & \langle \tilde{B}_z \tilde{B}_x^* \rangle \\ \langle \tilde{B}_x \tilde{B}_y^* \rangle & \langle \tilde{B}_y \tilde{B}_y^* \rangle & \langle \tilde{B}_z \tilde{B}_y^* \rangle \\ \langle \tilde{B}_x \tilde{B}_z^* \rangle & \langle \tilde{B}_y \tilde{B}_z^* \rangle & \langle \tilde{B}_z \tilde{B}_z^* \rangle \end{pmatrix}. \tag{3.7}$$

Hereafter the factor $1/T$ is omitted and we simply write the CSD matrix as $\mathsf{R} = \left\langle \tilde{\mathbf{B}} \tilde{\mathbf{B}}^\dagger \right\rangle$. The diagonal elements R_{xx}, R_{yy}, and R_{zz} are real numbers and represent the wave power in the respective components. The trace of the matrix gives therefore the total wave power as a function of the frequency:

$$P = \langle \tilde{B}_x \tilde{B}_x^* \rangle + \langle \tilde{B}_y \tilde{B}_y^* \rangle + \langle \tilde{B}_z \tilde{B}_z^* \rangle. \tag{3.8}$$

The off-diagonal elements are correlations between different fluctuation components and are Hermitian symmetric, e.g., $R_{xy} = R_{yx}^*$. These elements contain information about wave polarization such as linear or circular polarization, or right-handed or left-handed circular sense. We introduce two useful coordinate systems for analyzing wave and turbulence using the CSD matrix: One is the Mean-Field-Aligned system (MFA) and the other is the Minimum-Variance system (MV).

Mean-Field-Aligned System

In the MFA system the z-axis is set to the direction of the mean or large-scale magnetic field, and the x-y-plane is perpendicular to it. The orientation of the x- or y-direction needs another reference direction, for example, the Earth-to-Sun direction. The CSD matrix in the MFA system is given as

$$
\mathsf{R}_{\mathrm{MFA}} = \begin{pmatrix} \langle \tilde{B}_{\perp 1} \tilde{B}_{\perp 1}^* \rangle & \langle \tilde{B}_{\perp 2} \tilde{B}_{\perp 1}^* \rangle & \langle \tilde{B}_{\parallel} \tilde{B}_{\perp 1}^* \rangle \\ \langle \tilde{B}_{\perp 1} \tilde{B}_{\perp 2}^* \rangle & \langle \tilde{B}_{\perp 2} \tilde{B}_{\perp 2}^* \rangle & \langle \tilde{B}_{\parallel} \tilde{B}_{\perp 2}^* \rangle \\ \langle \tilde{B}_{\perp 1} \tilde{B}_{\parallel}^* \rangle & \langle \tilde{B}_{\perp 2} \tilde{B}_{\parallel}^* \rangle & \langle \tilde{B}_{\parallel} \tilde{B}_{\parallel}^* \rangle \end{pmatrix} .
\tag{3.9}
$$

The power of the perpendicular fluctuating component (incompressible sense) is $P_\perp = \langle \tilde{B}_{\perp 1} \tilde{B}_{\perp 1}^* \rangle + \langle \tilde{B}_{\perp 2} \tilde{B}_{\perp 2}^* \rangle$, whereas the power in the parallel fluctuation (compressible sense) is $P_\parallel = \langle \tilde{B}_\parallel \tilde{B}_\parallel^* \rangle$. The "$\perp 1$"component may be chosen as the maximum variance direction, the direction from the Earth to the Sun, or the flow direction in the perpendicular plane.

Polarization around the mean magnetic field (or the z-axis) can be determined from $\mathsf{R}_{\mathrm{MFA}}$. We model the fluctuation as an elliptically polarized wave around the mean field,

$$
\delta \mathbf{B} = \begin{pmatrix} a\, e^{i\omega t} \\ b\, e^{i(\omega t - \pi/2)} \\ 0 \end{pmatrix} ,
\tag{3.10}
$$

with the amplitudes under condition $|a| < |b|$. This gives the model CSD matrix as

$$
\mathsf{R}_{\mathrm{model}} = \begin{pmatrix} a^2 & iab & 0 \\ -iab & b^2 & 0 \\ 0 & 0 & 0 \end{pmatrix} .
\tag{3.11}
$$

We define the concept of the wave magnetic field ellipticity in the fashion of plasma physics, according to Stix [12] and Gary [13], that is the sense of field rotation around the direction of the static magnetic field when seen at a positive frequency,

$$
\varepsilon = \frac{b}{a} = \tan \phi .
\tag{3.12}
$$

Note that in optics and quantum mechanics, the convention is polarization with respect to the wave propagation vector $\mathbf{k}$ [14]. Since $|a| < |b|$, ellipticity varies between -1 and $+1$, in other words, $|\varepsilon| \leq 1$. The ellipticity angle ϕ can be determined by the measurement of the sub-matrix of R_{MFA} in the x-y-plane [15, 16],

$$\sin 2\phi = \frac{2 \, \text{Im} \, \left(R'_{\perp 1 \perp 2}\right)}{\left[(\text{tr}R')^2 - 4\det\left(R'\right)\right]^{1/2}},\tag{3.13}$$

where

$$R' = \begin{pmatrix} \langle \tilde{B}_{\perp 1} \tilde{B}^*_{\perp 1} \rangle & \langle \tilde{B}_{\perp 2} \tilde{B}^*_{\perp 1} \rangle \\ \langle \tilde{B}_{\perp 1} \tilde{B}^*_{\perp 2} \rangle & \langle \tilde{B}_{\perp 2} \tilde{B}^*_{\perp 2} \rangle \end{pmatrix}.\tag{3.14}$$

The meaning of ellipticity in our definition is as follows: $\varepsilon > 0$ at a positive frequency means that wave polarization is in the right-hand rotation sense (the same sense as electron gyration) when viewing into the direction of the mean magnetic field, independent of wave propagation direction. $\varepsilon = +1$ means that the wave is right-hand circularly polarized in the temporal sense. $\varepsilon = 0$ means that the wave is linearly polarized. $\varepsilon < 0$ means left-hand polarization that has the same sense as ion gyration about the mean magnetic field [12, 13]. It should not be confused that geophysical applications ([15] and [16]) follow the optics definition of polarization, but their method to calculate the ellipticity can be applied to the plasma physics definition of polarization as in Eqs. (3.12) and (3.13).

Minimum Variance System

The CSD matrix is a Hermitian matrix and can be diagonalized using a suitable unitary matrix. In other words, we find the directions of maximum, intermediate, and minimum variance that are mutually orthogonal, and rotate the matrix into the eigenvector system.

$$R_{MV} = \begin{pmatrix} \lambda_1 & 0 & 0 \\ 0 & \lambda_2 & 0 \\ 0 & 0 & \lambda_3 \end{pmatrix},\tag{3.15}$$

where $\lambda_1 \geq \lambda_2 \geq \lambda_3$ are the three eigenvalues of the CSD matrix and e_1, e_2, and e_3 are the eigenvectors associated with the eigenvalues. For a divergence-free field like the magnetic field, the wave vector direction is perpendicular to the fluctuation vector. This can be seen in the equation $\nabla \cdot \mathbf{B} = 0$ when replacing the ∇-operator by $i\mathbf{k}$ (plane wave assumption) and splitting the field into the constant field and fluctuating field, $\mathbf{B} = \mathbf{B}_0 + \delta\mathbf{B}$,

$$i\mathbf{k} \cdot \delta\mathbf{B} = 0.\tag{3.16}$$

Therefore the eigenvector associated with the smallest eigenvalue, e_3, is aligned with the wave vector direction with 180° ambiguity. This is the minimum variance analysis developed by Sonnerup and Cahill [17] to determine the wave propagation direction as well as the discontinuity normal direction using single-point measurements. Note that the method works only for elliptically or circularly polarized waves, and not for linearly polarized fluctuations, since the analysis is to determine the polarization plane. If only one eigenvalue is very large and the other two eigenvalues are small, the fluctuation is one-dimensional (linear polarization). If two eigenvalues are comparably large and the other is small, the fluctuation is almost circular and one can determine the wave vector direction. If all the three eigenvalues are of the same order, the fluctuation is isotropic in three dimensions, and can be interpreted either many waves simultaneously present at one frequency, or noise. Wave polarization (ellipticity) can also be investigated in the perpendicular plane to the wave propagation direction.

Minimum variance analysis is applied to many observations in geophysics and space physics. On the other hand, some fundamental properties of waves are still missing because of the limitations of single-point measurements. For example, wave propagation speed, frequency in the flow rest frame (that needs Doppler correction), and wavelength cannot be determined with single-point measurements without assuming a specific wave mode, i.e., dispersion relation.

3.2.1.1 Multi-Point Methods

It is of course ideal to have as many properly positioned spacecraft available as possible to Fourier transform observed fluctuations from the spatial coordinates into the wave numbers. The four measurement points of Cluster, from this point of view, are too few for performing the Fourier transform into a broad range of wave numbers, but a mathematical method has been proposed to estimate the energy distribution in the four-dimensional frequency-wave vector domain using four-point measurements only. The idea to use multiple spacecraft as a plasma wave array experiment was proposed before the concept of the Cluster mission was developed [18]. The idea of multi-spacecraft data analysis for studying waves and turbulence was further developed and adapted for the Cluster mission under the concept of projection [6, 7, 19, 20], which had earlier been used in seismic wave studies [21, 22]. A number of analysis methods were developed for Cluster data along with the projection methods, and are summarized in various literature, e.g., [23] and [24, 25].

3.2.1.2 Projection Methods

We define the state vector for the L-point measurements as

$$\mathbf{S}(\omega) = \begin{pmatrix} S(\omega, \mathbf{r}_1) \\ S(\omega, \mathbf{r}_2) \\ \vdots \\ S(\omega, \mathbf{r}_L) \end{pmatrix}. \tag{3.17}$$

Here each sensor measures a scalar quantity S at position $\mathbf{r}_i$. The field S is already transformed from the time series into the frequency domain, and is a function of the angular frequency ω and the position of the i-th sensor $\mathbf{r}_i$ (we omit the tilde symbol for the Fourier transform hereafter). Projection means extracting some useful information about the spatial structure from this vector, in particlar, the amplitude as a function of frequency and wave vector. Using the concept of projection the state vector is reduced to a scalar by taking a dot product with a suitable weight vector. We measure a scalar field using a sensor array and construct the state vector in the frequency domain; and the state vector is reduced to a scalar (which is the wave amplitude in complex number, retaining the phase information) in the frequency and wave vector domain,

$$S(\omega, \mathbf{k}) = \mathbf{w}^\dagger(\omega, \mathbf{k}) \cdot \mathbf{S}(\omega), \tag{3.18}$$

where $\mathbf{w}^\dagger(\omega, \mathbf{k})$ denotes the weight vector (the dagger means the operation of Hermitian conjugate). One may also estimate the wave power in the frequency and wave vector domain by taking square of the projected quantity:

$$P(\omega, \mathbf{k}) = |S(\omega, \mathbf{k})|^2 = \mathbf{w}^\dagger(\omega, \mathbf{k})\mathsf{R}(\omega)\mathbf{w}(\omega, \mathbf{k}) \tag{3.19}$$

with the L-by-L CSD matrix of the state vector $\mathsf{R}(\omega)$:

$$\mathsf{H}(\omega) = \left\langle \mathbf{S}(\omega)\mathbf{S}^\dagger(\omega) \right\rangle. \tag{3.20}$$

The task is now to find the weight vector in a suitable form describing the wave power associated with the parametric wave vector $\mathbf{k}$. We introduce two different projection methods: beam-former projection and Capon's minimum variance projection.

3.2.1.3 Beam-Former Projection

Let us define the steering vector that describes a plane wave pattern at each sensor position characterized by the wave vector $\mathbf{k}$:

$$\mathbf{h}(\mathbf{k}) = \begin{pmatrix} e^{i\mathbf{k}\cdot\mathbf{r}_1} \\ e^{i\mathbf{k}\cdot\mathbf{r}_2} \\ \vdots \\ e^{i\mathbf{k}\cdot\mathbf{r}_L} \end{pmatrix}. \tag{3.21}$$

The beam-former projection uses the steering vector as the weight vector,

$$\mathbf{w} = \mathbf{h}, \tag{3.22}$$

and the wave power is estimated as

$$P_{\mathrm{BF}} = \mathbf{h}^\dagger \mathsf{R} \mathbf{h}. \tag{3.23}$$

Fig. 3.4 Beam-former and
Capon spectra for synthetic
data containing one signal
wave at $k = 0.021$ rad/km
and noise

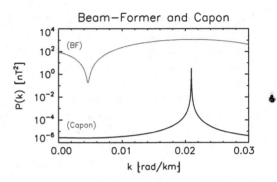

An example of the wave number spectrum reconstructed using the beam-former technique is displayed in Fig. 3.4. For the reconstruction test, synthetic data consisting of a single wave and noise were used. The generated data were sampled at four distinct spatial positions (representing four sensors) as time series data, and then the wave number spectrum is estimated from the multi-point time series data. The spectrum shows a peak at the signal wave number, that one used in generating a wave in the synthetic data. But the spectrum is very broad and the identified peak is not very reliable because the spectrum turns out to be very flat, dropping at relatively small wave numbers, which is much different from the true spectrum of the synthetic data.

The reason for the flat spectral curve from the beam-former projection is the fact that the number of sensors is too small. In fact, if a large number of sensors were available, the beam-former projection reasonably reproduces the Fourier series expansion of the data. If the number of sensors is too small, one obtain non-vanishing correlation with data not only at the signal wave number but also at other wave numbers. This phenomenon can be referred to as the cross-talk in the wave number domain for estimating the power spectrum. We therefore need to devise a method in which the cross-talk is minimized.

3.2.1.4 Capon's Minimum Variance Projection

To reduce the high background level of the beam-former spectrum and to make the spectrum have a sharp peak at the right signal source, Capon [21] proposed the method of minimum variance projection. Consider to minimize the power while keeping the fluctuation amplitude at the given wave vector **k** unchanged, minimizing only the cross-talk effect. This problem can be formulated as an optimization problem under a constraint:

$$\text{minimize} \quad \mathbf{w}^{\dagger}\mathsf{R}\mathbf{w} \quad \text{subject to} \quad \mathbf{w}^{\dagger} \cdot \mathbf{h} = 1$$

or

$$\delta \left[\mathbf{w}^\dagger \mathsf{R}\mathbf{w} - \lambda \left(\mathbf{w}^\dagger \cdot \mathbf{h} - 1 \right) \right] = 0, \tag{3.24}$$

with the Lagrangian multiplier λ. Capon [21] obtained the analytical expression for the projection weight as:

$$\mathbf{w} = \frac{\mathsf{R}^{-1}\mathbf{h}}{\mathbf{h}^\dagger \mathsf{R}^{-1}\mathbf{h}}. \tag{3.25}$$

The projected power is therefore

$$P_{\mathrm{C}}(\omega, \mathbf{k}) = \frac{1}{\mathbf{h}^\dagger(\mathbf{k})\mathsf{R}^{-1}(\omega)\mathbf{h}(\mathbf{k})}. \tag{3.26}$$

See, for example, Haykin [26] for the derivation. It is worthwhile to note that Capon's projection or weight vector is determined not only by the steering vector but also by the measurement itself through the state vector. Also, the matrix R must be invertible. An example of the spectrum estimated by Capon's method is presented in Fig. 3.4, too. The spectrum shows a much clearer peak at the signal wave number and the background level is significantly reduced.

3.2.1.5 Wave Telescope/k-Filtering Technique

The projection method can be generalized to measurements of a vector quantity such as the magnetic field. In this case the state vector has $3L$ elements (3 components of the vector measured by L sensors):

$$\mathsf{S}(\omega) = \begin{pmatrix} \mathbf{B}_1(\omega) \\ \mathbf{B}_2(\omega) \\ \vdots \\ \mathbf{B}_4(\omega) \end{pmatrix}, \tag{3.27}$$

and the generalized CSD matrix is

$$\mathsf{R}(\omega) = \frac{1}{T} \langle \mathsf{S}(\omega)\mathsf{S}^\dagger(\omega) \rangle. \tag{3.28}$$

The CSD matrix is a $3L \times 3L$ matrix and depends on the frequency. After the projection, the CSD matrix is reduced to a 3×3 matrix as function of frequency and wave vector. The steering vector becomes a $3L \times 3$ matrix and is given as

$$\mathsf{H}(\mathbf{k}) = \begin{pmatrix} \mathsf{I} \; e^{i\mathbf{k}\cdot\mathbf{r}_1} \\ \mathsf{I} \; e^{i\mathbf{k}\cdot\mathbf{r}_2} \\ \mathsf{I} \; e^{i\mathbf{k}\cdot\mathbf{r}_3} \\ \mathsf{I} \; e^{i\mathbf{k}\cdot\mathbf{r}_4} \end{pmatrix}, \tag{3.29}$$

where I denotes the 3×3 unit matrix. Capon's projection can be generalized to projection into a vector, (B_x, B_y, B_z) as a function of frequency and wave vector. The weight vector is then generalized to the weight matrix. It can be expressed formally in the same fashion as the scalar projection case.

$$W(\omega, \mathbf{k}) = \frac{R^{-1}(\omega)H(\mathbf{k})}{H^{\dagger}(\mathbf{k})R^{-1}(\omega)H(\mathbf{k})} \tag{3.30}$$

under the unit gain constraint

$$W^{\dagger}H = I. \tag{3.31}$$

The power matrix is therefore obtained as

$$P(\omega, \mathbf{k}) = \left[H^{\dagger}(\mathbf{k})R^{-1}(\omega)H(\mathbf{k})\right]^{-1}. \tag{3.32}$$

The power matrix can be interpreted as a CSD matrix in the frequency and wave vector domain. Its diagonal elements represent the wave power in the x, y, and z components and the off-diagonal elements represent wave polarization or helicity. The trace of the matrix gives the total fluctuation power.

Furthermore, one may impose an additional constraint that the fluctuating field satisfies the divergence-free condition [20]. This can be expressed as

$$\mathbf{k} \cdot W^{\dagger}S = 0. \tag{3.33}$$

This means that one may insert a 3×3 matrix V such that S is replaced by VS:

$$V = I + \frac{\mathbf{k}\mathbf{k}}{k^2}. \tag{3.34}$$

The matrix V can be directly incorporated in Capon's projection estimator and we obtain the weight matrix as [20]

$$W(\omega, \mathbf{k}) = R^{-1}(\omega)H(\mathbf{k})V(\mathbf{k})\left[V^{\dagger}(\mathbf{k})H^{\dagger}(\mathbf{k})R^{-1}(\omega)H(\mathbf{k})V(\mathbf{k})\right]^{-1} \tag{3.35}$$

and the 3×3 power matrix:

$$P_{WT}(\omega, \mathbf{k}) = \left[V^{\dagger}(\mathbf{k})H^{\dagger}(\mathbf{k})R^{-1}(\omega)H(\mathbf{k})V(\mathbf{k})\right]^{-1}, \tag{3.36}$$

and its trace gives the total fluctuation power:

$$P_{WT}(\omega, \mathbf{k}) = \text{tr}\left(\left[V^{\dagger}(\mathbf{k})H^{\dagger}(\mathbf{k})R^{-1}(\omega)H(\mathbf{k})V(\mathbf{k})\right]^{-1}\right). \tag{3.37}$$

Estimating the power in the wave vector domain using the three matrices R, H, and V is called the wave telescope technique or the k-filtering [7, 19, 20, 28]. It provides the means to determine the energy distribution in the four-dimensional frequency and

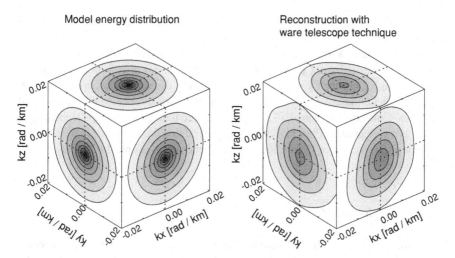

Fig. 3.5 Comparison of energy distribution in the three-dimensional wave vector domain. Upper panel displays the model distribution from which synthetic data are generated and sampled at four different positions in the spatial coordinate. Lower panels displays the energy distribution reconstructed using the wave telescope technique. Adapted from the numerical test of the wave telescope technique in [10]

wave vector domain using Cluster data. Further extensions of the filtering technique were developed by Motschmann et al. [29], coping with wave polarization and plasma mode recognition.

The projection vector $\mathbf{w}$ and the projection matrix $\mathbf{W}$ are the dimensionless operators and they do not change the units of the spectrum after the projection, i.e., the same dimension as the CSD matrix (squared amplitude per frequency). A suitable procedure is needed to properly interpret the results of Capon's spectrum such as integration over frequency or division by the wave number interval. Figure 3.5 shows comparison of the energy distributions in the three-dimensional wave vector domain: model distribution (left panel) and reconstruction using the wave telescope technique from synthetic time series data sampled at four discrete spatial points.

There are also limitations in the wave telescope technique. From the viewpoint of calculation, the spectral estimator of the wave telescope technique works for arbitrary wave vectors. In practice, tetrahedral configuration imposes the upper limit and the lower limit. The upper limit of the wave number is determined by the spacecraft separation (Nyquist wave number), while the lower limit is determined by the cognition of large-scale wavelength. Sahraoui et al. [30] evaluated that 1/50 of the maximum wave number is reasonable for choosing the lower limit of the wave number. The wave telescope does not require any knowledge on the number of signals. In the eigenvector method (explained later), on the other hand, one has to know the number of signals. The wave telescope technique was extensively tested using synthetic data [7, 19, 20]. It was also discussed by Glassmeier et al. [7] that the mean field should be uniform in the collected data to apply the wave telescope technique.

An example of the three-dimensional energy distribution in the wave vector domain reconstructed by the wave telescope technique is displayed in Fig. 3.5 together with the model distribution used in the test data [10]. Although the reconstructed distribution is not exactly the same as the model distribution, the overall structure can be reasonably well reconstructed using the wave telescope technique. Capon's projection method is valid not only for plane waves but also for other spatial structures. Constantinescu et al. [31, 32] generalized the wave telescope technique to spherical wave patterns. Plaschke et al. [33] generalized the technique to a field-line-resonant phase pattern of ULF (ultra-low-frequency) pulsations of the geomagnetic field.

3.2.1.6 Energies and Helicities

The wave telescope technique provides the 3×3 matrix with each element representing correlation among B_x, B_y, and B_z components of fluctuations in the frequency and wave vector domain. While the diagonal elements of the CSD matrix represent self-correlation and thus the wave power, the off-diagonal elements are cross-correlation and contain information about the spatial rotation sense of the fluctuation, i.e., the helicity.

The energy distribution or the spectrum in the three-dimensional wave vector domain can be obtained by (1) integrating the power over frequencies and then (2) dividing the power by the wave number interval Δk for transforming the units from the spectral power into the spectral power density. The energy distribution (in units of squared amplitude) in the three-dimensional wave vector domain is obtained as

$$P(\mathbf{k}) = \int P_{\mathrm{WT}}(\omega, \mathbf{k}) d\omega, \tag{3.38}$$

and the energy density distribution in the three-dimensional wave vector domain is

$$P'(\mathbf{k}) = \frac{1}{\Delta^3 k} \int P_{\mathrm{WT}}(\omega, \mathbf{k}) d\omega, \tag{3.39}$$

where $\Delta^3 k$ denotes the three-dimensional volume of the gridded wave vector domain. This is closely related to the magnetic pressure (or magnetic energy density). Assuming that the proper factor and integration are used for the dimensional matching of the wave telescope spectrum, one may write down the magnetic energy in the wave vector domain as

$$E^M = P_{xx} + P_{yy} + P_{zz}. \tag{3.40}$$

The off-diagonal parts of the CSD matrix can also be used to describe further physical quantities. The current helicity density is defined as

$$\mathbf{j} \cdot \mathbf{B} = \frac{1}{\mu_0} (\nabla \times \mathbf{B}) \cdot \mathbf{B} \tag{3.41}$$

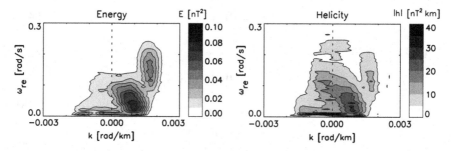

Fig. 3.6 Energy and helicity density in the frequency-wavenumber domain for shock-upstream (foreshock) fluctuations from Cluster data [34]

and this can be expressed in the Fourier domain by replacing the ∇-operator by $i\mathbf{k}$ (plane wave assumption) as

$$\mathbf{j}^\dagger \cdot \mathbf{B} = -\frac{i}{\mu_0}(\mathbf{k} \times \mathbf{B}) \cdot \mathbf{B} \tag{3.42}$$

$$= -\frac{i}{\mu_0}\left[k_x\left(B_y^* B_z - B_z^* B_y\right) + k_y\left(B_z^* B_x - B_x^* B_z\right) \right.$$
$$\left. + k_z\left(B_x^* B_y - B_y^* B_x\right)\right]. \tag{3.43}$$

We take the statistical average of $\mathbf{j}^\dagger \cdot \mathbf{B}$ (either temporal or spatial averaging) and use the wave telescope estimator for the cross-correlation terms $P_{ij} = \langle B_i^* B_j \rangle$, which gives the current helicity density in the form

$$\langle \mathbf{j}^\dagger \cdot \mathbf{B} \rangle = -\frac{i}{\mu_0}\left[k_x(P_{yz} - P_{zy}) + k_y(P_{zx} - P_{xz}) + k_z(P_{xy} - P_{yx})\right]. \tag{3.44}$$

In a similar fashion, the magnetic helicity density can also be obtained as

$$\langle \mathbf{A}^\dagger \cdot \mathbf{B} \rangle = -\frac{i}{k^2}\left[k_x(P_{yz} - P_{zy}) + k_y(P_{zx} - P_{xz}) + k_z(P_{xy} - P_{yx})\right]. \tag{3.45}$$

This can be checked easily under the Coulomb gauge $\nabla \cdot \mathbf{A} = 0$ [34]. Examples of energy and helicity density in the frequency-wavenumber domain are presented in Fig. 3.6

The method can be applied to the flow velocity vector if data are available from multi-point measurements. The wave telescope technique for the flow velocity uses the form of Eq. (3.32), noting that the flow in general may be compressible ($\nabla \cdot \mathbf{u} \neq 0$):

$$\mathsf{P}_u(\omega, \mathbf{k}) = \left[\mathsf{H}^\dagger(\mathbf{k})\mathsf{R}_u^{-1}(\omega)\mathsf{H}(\mathbf{k})\right]^{-1}, \tag{3.46}$$

where R_u is the CSD matrix of the flow velocity:

$$R_u = \left\langle S_u S_u^\dagger \right\rangle, \tag{3.47}$$

with the state vector

$$S_u(\omega) = \begin{pmatrix} \mathbf{u}(\omega, \mathbf{r}_1) \\ \mathbf{u}(\omega, \mathbf{r}_2) \\ \vdots \\ \mathbf{u}(\omega, \mathbf{r}_L) \end{pmatrix}. \tag{3.48}$$

The kinetic energy (per unit mass) is therefore given as the trace of the matrix P_u after integration over frequencies,

$$E^K = P_{u,xx} + P_{u,yy} + P_{u,zz}, \tag{3.49}$$

and the kinetic helicity density $h^K = \mathbf{u} \cdot \mathbf{\Omega}$ is given as

$$h^K = \left\langle \mathbf{u}^\dagger \cdot \mathbf{\Omega} \right\rangle \tag{3.50}$$

$$= i \left[k_x (P_{u,yz} - P_{u,zy}) + k_y (P_{u,zx} - P_{u,xz}) + k_z (P_{u,xy} - P_{u,yx}) \right], \tag{3.51}$$

where we used the definition of the vorticity in the Fourier domain as:

$$\mathbf{\Omega} = i \mathbf{k} \times \mathbf{u}. \tag{3.52}$$

It is therefore possible to determine the enstrophy (squared vorticity),

$$\Omega^2 = |\mathbf{k} \times \mathbf{u}|^2. \tag{3.53}$$

In addition, one may determine compressibility of the flow. The concept of dilatation is defined as the divergence of the flow,

$$d = \nabla \cdot \mathbf{u}. \tag{3.54}$$

We interpret the dilation in the Fourier domain and it is estimated from the measurement as

$$\tilde{d} = i\mathbf{k} \cdot \tilde{\mathbf{u}} = i\mathbf{k} \cdot W_u S_u, \tag{3.55}$$

where W_u is the weight matrix for the flow velocity:

$$W_u(\omega, \mathbf{k}) = \frac{R_u^{-1}(\omega) H(\mathbf{k})}{H^\dagger(\mathbf{k}) R_u^{-1}(\omega) H(\mathbf{k})}. \tag{3.56}$$

Finally, the coupling between the flow velocity and the magnetic field gives the cross helicity density,

$$h^C = \langle \tilde{\mathbf{u}}^\dagger \cdot \tilde{\mathbf{b}} \rangle \tag{3.57}$$

$$= (\mathbf{W}_u \mathbf{S}_u)^\dagger \cdot \mathbf{W}_b \mathbf{S}_b, \tag{3.58}$$

or, using the trace of the CSD matrix,

$$h^C = \text{tr} \left\langle \tilde{\mathbf{u}} \cdot \tilde{\mathbf{b}}^\dagger \right\rangle \tag{3.59}$$

$$= \text{tr} \left(\mathbf{W}_u \langle \mathbf{S}_u \mathbf{S}_b \rangle \mathbf{W}_b \right), \tag{3.60}$$

where $\mathbf{W}_b$ is given by Eq. (3.35).

3.2.1.7 Higher Order Moments

Using the wave telescope technique it is also possible to evaluate higher order moments which are useful quantities to study wave-wave interactions. Third order moments or three-point correlations can be determined in the frequency and wave vector domain, which are a useful tool to evaluate at which combinations of frequencies or wave vectors the waves are interacting with one another. Three-wave processes are characterized by the resonance condition (or conservation) of frequencies and wave vectors, described as $\omega'' = \omega \pm \omega'$ and $\mathbf{k}'' = \mathbf{k} \pm \mathbf{k}'$. For example, one of the physically relevant third order moments is given as

$$C^\pm(\omega, \omega', \mathbf{k}, \mathbf{k}') = \langle b(\omega, \mathbf{k}) n(\omega, \mathbf{k}) b^*(\omega_\pm \omega', \mathbf{k} \pm \mathbf{k}') \rangle, \tag{3.61}$$

where we used, for simplicity, the scalar fields (magnetic field b and number density n),

$$b(\omega, \mathbf{k}) = \mathbf{w}_b^\dagger(\omega, \mathbf{k}) \cdot \mathbf{S}_b(\omega) \tag{3.62}$$

$$n(\omega, \mathbf{k}) = \mathbf{w}_n^\dagger(\omega, \mathbf{k}) \cdot \mathbf{S}_n(\omega), \tag{3.63}$$

and $\mathbf{S}_b(\omega)$ and $\mathbf{S}_n(\omega)$ are the state vectors for the scalar magnetic field and the number density, and $\mathbf{w}_b$ and $\mathbf{w}_n$ are Capon's weight vectors associated with the magnetic field and density, respectively. The meaning of the bispectrum is as follows. If three waves are in resonance, the sum of frequency and the sum of the wave vectors are conserved before and after the interaction (analogy to energy and momentum conservation of photons in quantum mechanics), and the bispectrum is non-zero. If three waves have frequencies and wave vectors (and initial phases, too!) that do not satisfy the resonant condition, the bispectrum is small or almost zero because of the statistical averaging. An example of the bispectrum in the wave vector domain is displayed in Fig. 3.7 for a synthetic data set containing three waves in resonance. In a similar fashion, the fourth or higher order moments can be determined in the frequency and wave vector domain

Fig. 3.7 Bispectrum in the wave number domain estimated for Cluster data in the foreshock region [35]

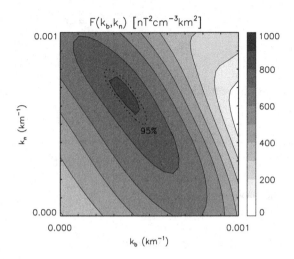

3.2.2 Eigenvalue-Based Methods

The CSD matrix is Hermitian symmetric and has eigenvalues and eigenvectors which also have information about spatial structure of waves. Eigenvectors of the CSD matrix can be used to construct an analysis tool. Here we discuss two applications of the eigenvector analysis of the CSD matrix: the wave surveyor technique [36] and the multi-point signal resonator technique [40]. The former provides a fast wave vector identification method and the latter provides a very-high-resolution spectrum in the wave number domain.

3.2.2.1 Wave Surveyor Technique

The wave surveyor technique [36] is a direct identification tool of wave dispersion relations in the sense that the wave vector is computed as a function of the frequency. The wave telescope technique, in contrast, provides the estimate of wave power in the frequency-wave vector domain and requires a peak search of the wave power in the three-dimensional wave vector domain for the dispersion analysis. The wave surveyor technique makes use of the eigenvector-decomposition of the CSD matrix, and it essentially computes the wave vector directly from the difference of wave phase measured at different sensors, assuming a propagating plane wave. The eigenvectors of the CSD matrix are given as a set of complex numbers and their phases θ_i corresponding to the i-th eigenvalue (in descending order) should be equal or close to the spatial wave phase in observer's frame (spacecraft frame). We therefore minimize the square-deviation between the ideal phase shift for a plane wave and the measured phase shift, i.e., minimize the following function

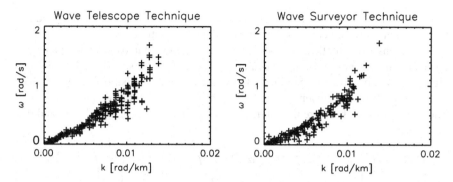

Fig. 3.8 Dispersion relation of foreshock waves measured by Cluster. *Left panel* displays the dispersion relation determined by the wave telescope technique [37], and *right panel* displays that determined by the wave surveyor technique [36]. Frequencies are in the flow rest frame

$$Q(\mathbf{k}, \phi) = \sum_{i=1}^{L} [\theta_i - \mathbf{k} \cdot \mathbf{r}_i - \phi]^2, \qquad (3.64)$$

with respect to the wave vector $\mathbf{k}$ and the initial phase ϕ. It was presented that the wave vector can be directly obtained from the eigenvector phases as [36]

$$\mathbf{k} = \left(\sum_i \mathbf{r}_i \mathbf{r}_i^t \right)^{-1} \sum_i \theta_i \mathbf{r}_i. \qquad (3.65)$$

Here $\mathbf{r}_i^t$ is the transposed vector of $\mathbf{r}_i$ at i-th sensor, and the sensor positions are measured from the center of the sensor array, i.e., they satisfy the condition $\sum_i \mathbf{r}_i = 0$.

In the case of four sensors ($L = 4$) like the Cluster mission, the solution can be explicitly given as a linear combination of the reciprocal vectors κ_i of the spacecraft positions,

$$\mathbf{k}(\omega) = \sum_i \theta_i \mathbf{k}_i, \qquad (3.66)$$

where $\mathbf{k}_i$ are the reciprocal vectors. This method directly determines the wave vector associated with the frequency ω by investigating the phase difference between the sensors and is called the wave surveyor technique [36]. The wave surveyor technique gives a very similar result in the dispersion relation analysis to that of the wave telescope technique (Fig. 3.8).

3.2.2.2 MUSIC Algorithm and MSR Technique

The MUSIC algorithm (MUltiple SIgnal Classification) is an application of eigenvalue-based spectrum. It suppresses the noise contribution considerably to

provide a sharp contrast between the signal and the noise-background in the spectrum
by investigating the eigenvector structure of the CSD matrix.

The eigenvectors of the CSD matrix are strictly orthogonal to one another and this
fact can be used for establishing an estimator of wave number spectrum. The MUSIC
algorithm was proposed by Schmidt [38] on the assumption that the measured data
contain signal and noise such that the CSD matrix can be decomposed into two parts.
The state vector is therefore interpreted as a combination of the signal term and the
noise term:

$$
\begin{pmatrix} S(\omega, \mathbf{r}_1) \\ S(\omega, \mathbf{r}_2) \\ \vdots \\ S(\omega, \mathbf{r}_L) \end{pmatrix} = \begin{pmatrix} e^{ik_1 \cdot r_1} & \cdots & e^{ik_M \cdot r_1} \\ \vdots & \ddots & \vdots \\ e^{ik_1 \cdot r_L} & \cdots & e^{ik_M \cdot r_L} \end{pmatrix} \begin{pmatrix} Q_1(\omega, k_1) \\ Q_2(\omega, k_2) \\ \vdots \\ Q_M(\omega, k_M) \end{pmatrix} + \begin{pmatrix} N_1(\omega) \\ N_2(\omega) \\ \vdots \\ N_L(\omega) \end{pmatrix},
$$

(3.67)

which we write symbolically as

$$
\mathbf{S} = \mathbf{AQ} + \mathbf{N}. \tag{3.68}
$$

Here $\mathbf{Q}$ represents the M true signals in the data and it is transmitted to the measurement of the state vector $\mathbf{S}$ through the matrix $\mathbf{A}$. $\mathbf{N}$ represents the noise at each sensor.
It can be shown that under this concept the CSD matrix is decomposed into the signal
term and the noise term as

$$
\mathbf{R} = \mathbf{A}\langle \mathbf{QQ}^\dagger \rangle \mathbf{A}^\dagger + \sigma^2 \mathbf{I}. \tag{3.69}
$$

The eigenvalues and eigenvectors of $\mathbf{R}$ are denoted by $\lambda_1 \geq \lambda_2 \geq \ldots \geq \lambda_L$
and $\mathbf{e}_1, \mathbf{e}_2, \ldots, \mathbf{e}_L$, respectively. For the noise part the eigenvalues are given
as $\lambda_{M+1} = \lambda_{M+2} = \ldots = \lambda_L = \sigma^2$. We split the eigenvectors of the CSD
matrix $\mathbf{R}$ into the signal subspace $\mathbf{E}_s = [\mathbf{e}_1, \mathbf{e}_2, \ldots, \mathbf{e}_M]$ and the noise subspace
$\mathbf{E}_n = [\mathbf{e}_{M+1}, \mathbf{e}_{M+2}, \ldots, \mathbf{e}_L]$.

The power estimation in the MUSIC method is given as

$$
P_{\text{MUSIC}}(\omega, \mathbf{k}) = \frac{1}{|\mathbf{h}^\dagger(\mathbf{k})\mathbf{E}_n(\omega)|^2} \tag{3.70}
$$

$$
= \frac{1}{\mathbf{h}^\dagger(\mathbf{k})\mathbf{E}_n(\omega)\mathbf{E}_n^\dagger(\omega)\mathbf{h}(\mathbf{k})}, \tag{3.71}
$$

which makes use of the orthogonality between the steering vector $\mathbf{h}(\mathbf{k}_i)$ ($i = 1, \ldots, M$) and the eigenvector for the noise part $\mathbf{e}_j$ ($j = M + 1, \ldots, L$):

$$
\mathbf{h}^\dagger(\mathbf{k}_i) \cdot \mathbf{e}_j = 0. \tag{3.72}
$$

The MUSIC spectrum is also expressed as

$$
P_{\text{MUSIC}}(\omega, \mathbf{k}) = \frac{1}{\mathbf{h}^\dagger(\mathbf{k})\mathbf{F}(\omega)\mathbf{L}\mathbf{F}^\dagger(\omega)\mathbf{h}(\mathbf{k})}, \tag{3.73}
$$

where F is the eigenvector matrix of R sorted after the magnitude of eigenvalues in the descending order:

$$F = [E_s E_n] = \begin{bmatrix} e_1 \ldots e_M \ e_{M+1} \ldots e_L \end{bmatrix}. \tag{3.74}$$

The matrix F is an arrangement of the eigenvectors of the CSD matrix, sorting the signal-associated eigenvectors on the left side in the matrix and the noise-associated eigenvectors on the right side. The matrix L is a diagonal matrix and defined as

$$L = \text{diag} \underbrace{(0, \ldots, 0,}_{M} \underbrace{1, \ldots, 1)}_{L-M}. \tag{3.75}$$

The MUSIC algorithm is based on finding the eigenvectors associated with noise that are orthogonal to the steering vector with the signal wave vector. The spectrum estimated by the MUSIC algorithm uses the product of the noise eigenvectors and the steering vectors and therefore the method gives the spectrum in the dimensionless unit. It should also be noted that the MUSIC algorithm requires that the number of signals must be known in the analysis to extract the set of the eigenvectors associated with noise. One method to determine the number of signals is to investigate the eigenvalue spectrum, a plot of the eigenvalues in descending order, to classify the eigenvalues into signals or noise as displayed [26].

The problem that the number of signal sources must be known in the MUSIC algorithm can be solved [39] by replacing the diagonal matrix L by M^{-n} with

$$M^{-n} = \text{diag} \left(\left(\frac{\lambda_1}{\lambda_L} \right)^{-n}, \left(\frac{\lambda_2}{\lambda_L} \right)^{-n}, \ldots, \left(\frac{\lambda_L}{\lambda_L} \right)^{-n} \right). \tag{3.76}$$

Here the power $-n$ is an adjustable parameter in the analysis that controls the asymptotic behavior the matrix M^{-n} that becomes L in the limit $n \to \infty$. In other words, replacing the L matrix by the M^{-n} matrix automatically selects the noise subspace of the CSD matrix R. It should be noted that the procedure of the matrix replacement by M^{-n} does not stem from a mathematical theory guaranteeing the better functionality of the technique, but it represents an intuitive picture of generalization of the L matrix to soften its sharp transition in the diagonal elements from zero to unity. Therefore, other extensions or generalizations are possible for the MUSIC algorithm. It was found that even a small number of n such as $n = 2$ can successfully reproduce the MUSIC spectrum without knowing the number of signal sources [39]. The spectrum using the Extended-MUSIC algorithm is therefore given as

$$P_{\text{EM}}(\omega, \mathbf{k}) = \frac{1}{\mathbf{h}^\dagger(\mathbf{k}) F(\omega) M^{-n} F^\dagger(\omega) \mathbf{h}(\mathbf{k})}. \tag{3.77}$$

The estimator P_{EM} can be used as a filter to Capon's spectrum in the following way. The MSR technique (Multi-point Signal Resonator [40]) makes use of Capon's estimator as well as the Extended-MUSIC estimator. The notion of the MSR is as follows. We use Capon's estimator and obtain the power spectrum that exhibits the

Fig. 3.9 Energy spectrum for Cluster data in the solar wind derived by the wave telescope and the MSR techniques [41]

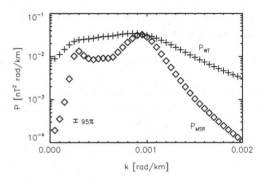

right value of the spectrum at the signal wave number, and we use additionally the Extended-MUSIC spectrum with a proper normalization as a dimensionless filter to make the signal-to-noise contrast of the Capon spectrum sharper. The power spectrum in the MSR technique is therefore given as

$$P_{\text{MSR}}(\omega, \mathbf{k}) = \frac{1}{P_{EM0}} P_{EM}(\omega, \mathbf{k}) P_C(\omega, \mathbf{k}) \tag{3.78}$$

$$= \frac{1}{P_{EM0}} \frac{1}{\mathbf{h}^{\dagger} \mathbf{F} \mathbf{M}^{-n} \mathbf{F}^{\dagger} \mathbf{h}} \frac{1}{\mathbf{h}^{\dagger} \mathbf{R}^{-1} \mathbf{h}}. \tag{3.79}$$

Here the factor P_{EM0} denotes the normalization factor. It is determined by the maximum value of the spectrum P_{EM}

$$P_{\text{EM0}} = \max(P_{EM}) \tag{3.80}$$

In the MSR technique $P_{EM}(\omega, \mathbf{k})/P_{EM0}$ serves as a filter that returns the value of almost unity at the signal wave numbers and almost zero values otherwise.

MSR technique inherits the merits of the MUSIC algorithm in resolving waves with closely separated wavelengths; Two signal waves that have close wave numbers to each other can be resolved better using the MSR technique, since Capon's spectrum exhibits a peak with a larger broadening although the cross-talk effect was minimized by incorporating Capon's minimum variance estimator. The peak on the left side (at the smaller wave number) appears as a hump in the Capon spectrum. The advantages of the MSR technique are therefore (1) much reduced background level and (2) high-resolution in the wave vector domain. The MSR technique can also be used for a measurement of a vector quantity such as the magnetic field. It is also possible to set the divergence-free condition as an additional constraint.

Figure 3.9 displays a comparison of the two spectra between MSR and wave telescope using Cluster data in the solar wind. The spectrum derived by the wave telescope (upper curve) is rather flat and exhibits a peak (local maximum in the energy distribution) at a wave number of about 9.5×10^{-4} rad/km, while the one derived by MSR (lower curve) clearly identifies two distinct peaks, one at the same wave number and the other at about 3.0×10^{-4} rad/km. This second peak appears in the wave telescope spectrum as only a spectral break. The difference in the two spectral curves represents the isotropic fluctuation in all 12 measured field components that cannot be interpreted as a plane wave. In the wave telescope all the eigenvalues of the CSD matrix are used in energy estimation, while in MSR/MUSIC only significant eigenvalues and eigenvectors are used on the assumption that small eigenvalues represent isotropic, random-phase noise.

References

1. Taylor, G.I.: The spectrum of turbulence. Proc. R. Soc. Lond. A **164**, 476–490 (1938)
2. Escoubet, C.P., Fehringer, M., Goldstein, M.: The Cluster mission. Ann. Geophys. **19**, 1197–1200 (2001)
3. Unser, M.: Sampling—50 years after Shannon. Proc. IEEE. **88**, 569–587 (2000)
4. Kirchner, J.W.: Aliasing in $1/f^\alpha$ noise spectra: origins, consequences, and remedies. Phys. Rev. E **71**, 066110 (2005)
5. Narita, Y., Glassmeier, K.-H.: Spatial aliasing and distortion of energy distribution in the wave vector domain under multi-spacecraft measurements. Ann. Geophys. **27**, 3031–3042 (2009)
6. Neubauer, F.M., Glassmeier, K.-H.: Use of an array of satellites as a wave telescope. J. Geophys. Res. **95**, 19115–19122 (1990)
7. Glassmeier, K.-H., Motschmann, U., Dunlop, M., Balogh, A., Acuña, M. H., Carr, C., Musmann, G., Fornaçon, K.-H., Schweda, K., Vogt, J., Georgescu, E., Buchert, S.: Cluster as a wave telescope—first results from the fluxgate magnetometer. Ann. Geophys. **19**, 1439–1447 (2001) Correction in 21, 1071 (2003)
8. Sahraoui F., Pinçon J.L., Belmont G., Rezeau L., Cornilleau-Wehrlin N., Robert P., Mellul L., Bosqued J. M., Balogh A., Canu P., Chanteur G.: ULF wave identification in the magnetosheath: the k-filtering technique applied to Cluster II data. J. Geophys. Res. **108**, SMP1-1, Cite ID 1335, (2003). doi:10.1029/2002JA009587 Correction in 109, A04222, doi10.1029/2004JA010469 (2004)
9. Tjulin, A., Pinçon, J.L., Sahraoui, F., André, M., Cornilleau-Wehrlin N.: The k-filtering technique applied to wave electric and magnetic field measurements from the Cluster satellites. J. Geophys. Res. **110**, A11224, (2005). doi:10.1029/2005JA011125
10. Narita, Y., Sahraoui, F., Goldstein, M.L., Glassmeier, K.-H.: Magnetic energy distribution in the four-dimensional frequency and wave vector domain in the solar wind. J. Geophys. Res. **115**, A04101 (2010). doi:10.1029/2009JA014742
11. Kittel, C.: Introduction to Solid State Physics . 7th edn. Wiley, New York (1996)
12. Stix, T.H.: Waves in Plasmas. 7th edn. Springer, New York (1992)
13. Gary, S.P.: Theory of Space Plasma Microinstabilities. 7th edn. Cambridge Atmospheric and Space Science Series, Cambridge (1993)
14. Born, M., Wolf, E.: Principles of Optics. 6th edn. Pergamon press, New York (1980)
15. Fowler, R.A., Kotick, B.J., Elliott, R.D.: Polarization analysis of natural and artificially induced geomagnetic micropulsations. J. Geophys. Res. **72**, 2871–2883 (1967)
16. Arthur, C.W., McPherron, R.L., Means, J.D.: A comparative study of three techniques for using the spectral matrix in wave analysis. Radio Sci. **11**, 833–845 (1976)
17. Sonnerup, B.U.Ö., Cahill, Jr. L.J.: Magnetopause structure and attitude from Explorer 12 observations. J. Geophys. Res. **72**, 171 (1967)
18. Musmann, G., Beinroth, H.J., Denskat, U., Hente, B., Theile, B., Neubauer, F.: Proposal for a plasma wave array experiment to be flown on the ESA Spacelab, ESA (1974)

19. Pinçon, J.L., Lefeuvre, F.: Local characterization of homogeneous turbulence in a space plasma from simultaneous measurement of field components at several points in space. J. Geophys. Res. **96**, 1789–1802 (1991)
20. Motschmann, U., Woodward, T.I., Glassmeier, K.-H., Southwood, D.J., Pinçon, J.L.: Wavelength and direction filtering by magnetic measurements at satellite arrays: generalized minimum variance analysis. J. Geophys. Res. **101**, 4961–4965 (1996)
21. Capon, J.: High resolution frequency-wavenumber spectrum analysis. Proc. IEEE **57**, 1408–1418 (1969)
22. Harjes, H.P., Henger, M.: Array-seismologie. Z. Geophysik **39**, 865–905 (1973)
23. Glassmeier, K.-H., Motschmann, U., Schmidt, R.: Proceedings of the Workshop On Data Analysis Tools. 6th edn. ESA Publications Division, ESA SP-371 (1995)
24. Paschmann, G., Daly, P.W. (eds.): Analysis Methods for Multi-Spacecraft Data. ISSI Scientific Report, SR-001, ISSI/ESA (1998)
25. Paschmann, G., Daly, P.W. (eds.): Multi-Spacecraft Analysis Methods Revisited. ISSI Scientific Report, SR-008, ISSI/ESA (2008)
26. Haykin, S.: Adaptive Filter Theory, Prentice Hall Information and System Science Series. 2nd edn. Prentice-Hall Inc., New Jersey (1991)
27. Narita, Y., Glassmeier, K.-H., Motschmann, U.: Wave vector analysis methods using multi-point measurements, Nonlin. Processes Geophys. **17**, 383–394 (2010)
28. Pinçon, J.-L., Motschmann, U.: Multi-spacecraft filtering: general framework. In: Paschmann, G., Daly, P.W. (eds.) Analysis Methods for Multi-Spacecraft Data, ISSI Scientific Report, SR-001, Chap. 3, pp. 65–78, International Space Science Institute, Berne (1998)
29. Motschmann, U., Glassmeier, K.-H., and Pinçon, J.-L.: Multi-spacecraft filtering: plasma mode recognition, analysis methods for multi-spacecraft data. In: Paschmann, G., Daly, P., (eds.) ISSI Scientific Report SR-001, Chap. 4, pp. 79–89, International Space Science Institute, Berne (1998)
30. Sahraoui, F., Belmont, G., Goldstein, M., Rezeau, L.: Limitations of multi-spacecraft data techniques in measuring wavenumber spectra of space plasma turbulence. J. Geophys. Res. **115**, A04206 (2010). doi:10.1029/2009JA014724
31. Constantinescu, O.D., Glassmeier, K.-H., Motschmann, U., Treumann, R.A., Fornaçon, K.-H., Fränz, M.: Plasma wave source location using CLUSTER as a spherical wave telescope. J. Geophys. Res. **111**, A09221 (2006). doi:10.1029/2005JA011550
32. Constantinescu, O.D., Glassmeier, K.-H., Décréau, P.M.E., Fränz, M., Fornaçon, K.-H.: Low frequency wave sources in the outer magnetosphere. Ann. Geophys. **25**, 2217–2228 (2007)
33. Plaschke, F., Glassmeier, K.H., Constantinescu, O.D., Mann, I.R., Milling, D.K., Motschmann, U., Rae, I.J.: Statistical analysis of ground based magnetic field measurements with the field line resonance detector. Ann. Geophys. **26**, 3477–3489 (2008)
34. Narita, Y., Kleindienst, G., Glassmeier, K.-H.: Evaluation of magnetic helicity density in the wave number domain using multi-point measurements in space. Ann. Geophys. **27**, 3967–3976 (2009)
35. Narita, Y., Glassmeier, K.-H., Décréau, P.M.E., Hada, T. Motschmann U., Nariyuki, Y.: Evaluation of bispectrum in the wave number domain based on multi-point measurements. Ann. Geophys. **11**, 3389–3393 (2008)
36. Vogt, J., Narita, Y., Constantinescu, O.D.: The wave surveyor technique for fast plasma wave detection in multi-spacecraft data. Ann. Geophys. **26**, 1699–1710 (2008)
37. Narita, Y., Glassmeier, K.-H., Fränz, M., Nariyuki, Y., Hada, T.: Observations of linear and nonlinear processes in the foreshock wave evolution. Nonlin. Processes Geophys. **14**, 361–371 (2007)
38. Schmidt, R.O.: Multiple emitter location and signal parameter estimation. IEEE Trans. Ant., Prop. AP- **34**, 276–280 (1986)
39. Choi, J., Song, I., Kim, H.M.: On estimating the direction of arrival when the number of signal sources is unknown. Signal Process. **34**, 193–205 (1993)

40. Narita, Y., Glassmeier, K.-H., Motschmann, U.: High-resolution wave number spectrum using multi-point measurements in space–the Multi-point Signal Resonator (MSR) technique. Ann. Geophys. **29**, 351–360 (2011)
41. Narita, Y., Gary, S.P., Saito, S., Glassmeier, K.-H., Motschmann, U.: Dispersion relation analysis of solar wind turbulence. Geophys. Res. Lett. **38**, L05101 (2011). doi:10.1029/2010GL046588

Chapter 4
Turbulence Properties in Space Plasma

4.1 Solar Wind Tubulence

4.1.1 Earlier Observations

Plasma and magnetic field fluctuations in the solar wind are believed to be in a fully-developed turbulent state. The primary reason for this is the fact that fluctuations appear to be random in time series and the frequency spectrum (or the energy spectrum in the frequency domain) of the fluctuating field shows a curve which is reminiscent of the spectrum for fluid turbulence. Early spacecraft observations in 1960s revealed that frequency spectra of interplanetary magnetic field fluctuations are characterized by a power-law resembling Kolmogorov's inertial-range spectrum for fluid turbulence [1]. Subsequent investigations have confirmed the power-law spectral curve and demonstrated that the spectral index of this inertial range is close to $-5/3$ for magnetic field and flow velocity [2–6]. Furthermore, the spectrum at higher frequencies exhibit a spectral break into a steeper slope (Fig. 4.1), which is reminiscent of the spectral curve in the dissipation range [5].

Kolmogorov's theory for fluid turbulence is based on several assumptions, e.g., fluctuations to be isotropic and incompressible, and dissipation due to particle collisions on the scale of mean free path. The solar wind is not an ordinary gas dynamic flow but a plasma flow and these assumptions cannot be justified any more: Plasma motion and magnetic field fluctuations influence each other due to conductivity; Various kinds of wave modes could be the carrier of energy cascade; The existence of large-scale magnetic field introduces anisotropy; and the plasma is collisionless in the sense that the mean free path is comparable to or larger than the system size (e.g., the heliosphere). Fluctuations are nearly incompressible in the sense that they appear mostly perpendicular to the mean or large-scale magnetic field direction. The solar wind serves as a natural laboratory for studying physical processes of plasma turbulence with implications to astrophysical applications. Plasma turbulence is a wide-spread phenomenon in astrophysical systems, e.g., accretion disks, interstellar

Y. Narita, *Plasma Turbulence in the Solar System*, SpringerBriefs in Physics, DOI: 10.1007/978-3-642-25667-7_4, © The Author(s) 2012

Fig. 4.1 Single-spacecraft
observation of frequency
spectrum of magnetic field
fluctuations in the solar wind
[5]

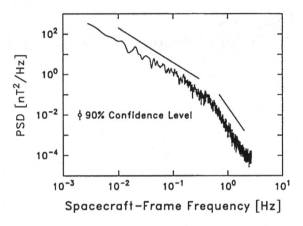

medium, stellar convection zones, so that our question is essential for understanding
the nature of this phenomenon. Understanding the reasons why the magnetic fluctuations in the solar wind exhibit the properties similar to fluid turbulence in spite
of collisionless plasma has been the subject of intense interest and studies for more
than 40 years.

Energy spectra of fluctuating magnetic field in the solar wind and their radial evolution with distance from the Sun have extensively been studied using in-situ measurements with Helios, Voyager, Ulysses, and many other spacecraft. The outward-
propagating waves dominate the spectrum in the high-speed solar wind close to the
Sun ($r \leq 0.3$ AU) and the spectrum is characterized by the power-law f^{-1} [4]
there. The shape of the spectrum evolves with increasing distance from the Sun. At
larger distances, $r \geq 1$ AU, the spectra of inward- and outward-propagating waves
merge into the Kolmogorov's inertial-range spectrum, $f^{-5/3}$. The fluctuations are
also intermittent, for the scaling exponents of higher order moments deviate strongly
from the linear law for Gaussian distribution [7]. Wave-particle interactions such as
cyclotron resonance were also found in the solar wind [8–10].

Anisotropy is one of the most interesting subjects not only in solar wind turbulence but also in studies of cosmic ray transport, and it was suggested that the
two-dimensional turbulence geometry must play a dominant role to account for long
mean free paths of cosmic rays [11, 12]. Using Taylor's hypothesis and projecting
magnetic field fluctuations to the axis parallel and perpendicular to the mean magnetic field, Matthaeus et al. [13] demonstrated that the fluctuations were anisotropic
and appeared to consist of two fluctuation populations: a population of planar fluctuations with large correlations perpendicular to the mean magnetic field and a second
population with large correlations parallel to the mean magnetic field (Fig. 4.2). In
other words, the solar wind fluctuations can be interpreted as a combination of two
components: a planar fluctuation geometry referred to as the slab geometry, representing wave packets piled up in the direction parallel to the mean magnetic field;
and a two-dimensional turbulence geometry such that the field lines are displaced
more or less randomly from one position to another without being bent. In the wave

Fig. 4.2 Two-dimensional correlation map in solar wind turbulence derived from single-spacecraft data using Taylor's frozen-in hypothesis and symmetries around the axis of the mean magnetic field and between parallel and anti-parallel directions to the mean field [13]

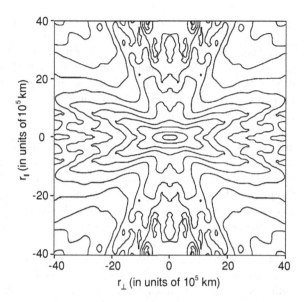

vector domain, the slab geometry is characterized by perpendicular wave vectors to the mean magnetic field, and the two-dimensional turbulence geometry by parallel wave vectors.

Due to limitations imposed by single point measurements, assumptions had to be made to derive the two-dimensional correlation contours in the spatial coordinate (which reflects the energy spectra in the wave vector domain by Fourier transform): Taylor's hypothesis relabeling frequencies into wave numbers along the flow; the energy spectrum to be axisymmetric about the mean or large-scale magnetic field; symmetry between parallel and anti-parallel direction to the large-scale field. Subsequent, extensive works confirmed the anisotropy with respect to the angle from the mean magnetic field [14, 15], but it was also found that high-speed and low-speed solar wind streams are dominated by different fluctuation geometries, the slab geometry for high-speed streams and the two-dimensional turbulence geometry for low-speed streams.

4.1.2 Three-dimensional Energy Spectrum

Multi-point measurements are very powerful in studying waves and turbulence in that they provide frequency and wave vector dependence of fluctuation properties. Although the Fourier transform into broad wave numbers is not possible, one still obtains a portion of the wave vector spectrum in three dimensions. Using Cluster data, the three-dimensional energy spectrum of magnetic field fluctuations in the solar wind was determined for the first time without employing Taylor's hypothesis nor symmetries in the wave vector domain. Rather than that, multi-point measurements offer the possibility to evaluate these assumptions.

The three-dimensional wave vector spectrum was derived from Cluster magnetic field data in the following steps. In the first step, the 12×12 cross spectral density (CSD) matrix is constructed in the frequency domain (in the spacecraft frame). Each element of the matrix represents cross-correlation of the magnetic field variation for different pairs of three components of the field measured at four spacecraft. Since Cluster performs measurements at spatially discrete points, the spatial aliasing effects must be taken into account in the wave vector domain as well as in the frequency domain because the Doppler effect relating spatial and temporal variations brings about aliasing in the frequency domain, too. The CSD matrix was measured inside the first Brillouin zone in the wave vector domain determined by the spacecraft separation distance. The limit of frequency (in the spacecraft frame) for the analysis is determined by investigating the periodic pattern of energy spectra in the frequency-wave vector domain (spatio-temporal aliasing).

In the second step, the CSD matrix is reduced into a 3×3 matrix (representing 3 components of the magnetic field) by projecting the matrix from the frequency domain into the four-dimensional frequency-wave vector domain using the wave telescope/k-filtering technique [16–19]. This technique is, as shown in the previous chapter, an generalization of Capon's minimum variance projection, and represents a parametric approach of estimating power spectra in the wave vector domain based on measurements only at several spatial points. In this projection method the filter coefficients are chosen so as to minimize the variance of the filter output under two constraints. The first is that the response of the filter is unity at the wave vector so as not to change the amplitude of waves by the projection, and the other is the divergence-free condition of the magnetic field. In the wave telescope technique, the wave vector serves as a parameter in the analysis, and the filter coefficients (projection weights) are determined not only by the wave vector but also by the measured CSD matrix itself for the purpose of reducing noise that comes from interference or cross-talk between different wave vectors in the spectral analysis.

The wave telescope technique can be regarded as a fitting procedure with propagating plane waves at various pairs of frequencies and wave vectors; this technique can determine the sign of wave vectors without use of cross helicity (which requires flow velocity data). Examples of distinguishing between forward and backward propagation directions using the wave telescope are displayed by Glassmeier et al. [19]. In the spacecraft frame the accessible wave vectors are symmetric with respect to changing the sign. However, when the measurement is performed in a stream such as in the solar wind the accessible frequencies and wave vectors in the plasma rest frame (co-moving frame with the flow) become asymmetric between the flow direction and the opposite direction to it, which needs to be taken into account.

In the third step, we obtain the fluctuation power by taking the trace of the reduced matrix as a function of frequencies and wave vectors in the spacecraft frame. The energy distribution is then transformed into the plasma rest frame by correcting for the Doppler shift. The mean flow velocity obtained by the electrostatic ion analyzer CIS-HIA on board Cluster [20] is used for the Doppler correction.

Fig. 4.3 Three-dimensional energy spectrum in the wave vector domain derived from Cluster data in the solar wind [21]. The 3D spectrum is projected onto three planes by summing over directions normal to these planes. Reprinted figure with permission from Narita et al., Phys. Rev. Lett., 104, 171101, 2010. Copyright 2010 by the American Physical Society

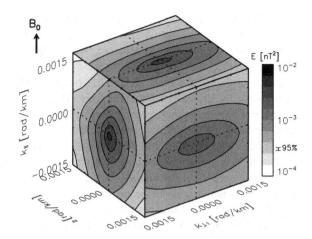

In the fourth step, the four-dimensional energy distribution is transformed into positive frequencies by changing the sign of frequencies and wave vectors for negative frequency components, and is then averaged over the rest-frame frequencies to obtain the three-dimensional wave vector spectrum. The frequency averaging is performed in the range in which the spectrum is symmetrically covered (or measured) between the flow direction and its opposite direction in the plasma rest frame.

The measurement of the wave vector spectrum using Cluster data provided (1) the direct evidence of the dominance of the perpendicular wave vector geometry (or the two-dimensional turbulence geometry) as well as (2) the lack of axisymmetry.

Figure 4.3 displays the three-dimensional energy distribution $E(\mathbf{k})$ in the MFA (Mean-Field-Aligned) coordinate system, spanned by the larger-scale magnetic field (assumed to be constant during the measurement) and the maximum power direction in the plane perpendicular to the mean field as $k_\parallel$ and $k_{\perp 1}$ components, respectively. The time interval of Cluster data represents a low-speed solar wind. The three-dimensional energy distribution was projected onto three planes by summing over the direction normal to them, $k_{\perp 1}$-$k_\parallel$ plane (right side panel), $k_{\perp 2}$-$k_\parallel$ plane (left side panel), and $k_{\perp 1}$-$k_{\perp 2}$ plane (top side panel). The distribution is overall symmetric with respect to changing the sign of the wave vector (reflection symmetry) but it is neither spherical nor axial. The primary extended structure appears in the $k_{\perp 1}$-$k_\parallel$ plane and the $k_{\perp 1}$-$k_{\perp 2}$ plane. The distribution also shows that axisymmetry fails around the mean magnetic field ($k_{\perp 1}$-$k_{\perp 2}$ plane), and this extended structure appeared to be almost perpendicular to the flow direction (in spite of the fact that the analysis was performed in the co-moving frame!). The secondary extended structure is present nearly parallel to the mean magnetic field direction ($k_{\perp 2}$-$k_\parallel$ plane). The small vertical line at the contour scale bar represents the 95%-confidence interval of the determined energy. The confidence interval is smaller than the contour scale and assures the small statistical error.

Quantitative analysis of the determined spectrum confirms that the perpendicular wave vector component contributes most to the fluctuation energy [21]. The ratio

$E(k_{\perp 1})$ to $E(k_\parallel)$ (each integrated over the other wave vector components) increases from almost unity at the lowest wave number to about 1.6 at the highest wave number, reflecting the primary extended structure. On the other hand, the ratio $E(k_{\perp 2})$ to $E(k_\parallel)$ decreases from almost unity to about 0.7 and this reflects the secondary extended structure parallel to the mean field. The spectral analysis was further extended to Cluster data for different tetrahedral sizes: 10,000, 1000 and 100 km, confirming the spectral anisotropy between parallel and perpendicular directions to the mean magnetic field and the asymmetry around the mean field on these scales [22].

It is interesting to note that single-spacecraft measurements indicated different result as to axisymmetry. Podesta [23] found, for example, that the frequency spectrum of magnetic field fluctuations in the high-speed solar wind is approximately azimuthally symmetric about the mean field. The solar wind has different phases such as high- and low-speed streams, plasma parameter beta, and the spectral anisotropy and asymmetry might be dependent on these phases.

The spectral anisotropy prefers the sense of energy cascade perpendicular to the mean magnetic field, and there are several possible explanations: (1) the anisotropy already exists in the solar corona and it is simply transported by the solar wind, (2) it develops in the interplanetary space by scattering of Alfvén waves, or (3) it develops due to radial expansion of the solar wind in the heliosphere. Radial expansion of the solar wind may influence solar wind turbulence. The expansion of the solar wind plasma causes stretching of eddies and anisotropy, and large-scale inhomogeneity must be present in the solar wind and turbulence can actively be excited along with its divergent flow. There may well be kinetic interaction with one of the fluctuation components.

4.1.3 Dispersion Relation

In multi-point measurements it is possible to study frequency-dependence of fluctuations in the co-moving frame, i.e., correcting for the Doppler shift. This allows one to determine the frequency-wave number diagram experimentally, to evaluate if plasma turbulence exhibits any particular dispersion relation or wave mode. The question on the role of wave modes in turbulence has been addressed in various numerical simulations, e.g., Parashar et al. [24], Svidzinski et al. [25], and Dmitruk and Matthaeus [26] finding only weak evidence for wave behavior.

While frequency spectra of magnetic field fluctuations in the solar wind are reminiscent of the picture of Kolmogorov's hydrodynamic turbulence (inertial range and dissipation range), the fluctuating fields do not always reach the level of the local mean or large-scale field. The concept of dispersion relation and associated normal modes, naïvely speaking, appears to be valid for sufficiently small amplitudes. In a turbulent medium, however, separation between the mean field and the fluctuating field is scale-dependent and nonlinearities can remain even for small amplitudes. One may therefore ask if there is any evidence for a dispersion relation in solar wind turbulence.

At very low frequencies (smaller than the proton gyro-frequency, Ω_p) and on large spatial scales (larger than the proton gyro-radius or inertial length), plasma turbulence should be described by magnetohydrodynamics (MHD), the fluid picture of plasma. Linear MHD theory predicts the existence of three normal modes: the incompressible shear Alfvén mode, the compressible 'slow' or ion acoustic mode, and the fast magnetosonic mode which exhibits both Alfvénic and acoustic properties. On sufficiently small scales, the MHD picture is no longer valid, and particle kinetic effects must be taken into account. Recent studies of solar wind turbulence showed that the magnetic field fluctuations exhibit two distinct spectral breaks and power-laws around $0.1 - 1$ and $10 - 100$ Hz in the spacecraft frame of reference [27], while Perri et al. [28] argue the spectral break around 0.1 Hz seems to be independent from proton gyro-frequency or -radius. The relative amplitude of electric to magnetic field fluctuation increases at higher frequencies [29], suggesting turbulence becomes more electrostatic in nature, but the existence of dispersion relation at high frequencies is still questionable [30].

The high-resolution wave vector analysis method, the MSR technique (Multipoint Signal Resonator), was applied to three time intervals of Cluster data in the solar wind for distinct tetrahedron scales: 10,000 (year 2006), 1000 (year 2005), and 100 km (year 2002) [31, 32]. The analysis of dispersion relation does not assume any relationship between frequency and wave vector but assumes that the local mean field is constant. The analysis procedure using MSR consists of two steps. First, wave vectors are determined that are associated with energy peaks at various frequencies in the ranges. Second, the set of frequencies and wave vectors are transformed into the plasma rest frame by correcting for the Doppler shift. Ion bulk velocity obtained by the ion spectrometer on board Cluster [20] was used for Doppler correction.

Two major results were obtained from the dispersion analysis: (1) nearly-perpendicular propagation on various spatial scales; (2) no clear dispersion relation accompanied by a weak signature of whistler mode waves.

Figure 4.4 top panel displays the distribution of propagation angles in the wave number domain normalized using the proton inertial length V_A / Ω_p. The diagram exhibits that wave vectors or propagation directions are quasi-perpendicular to the local mean magnetic field on these scales. The averaged angle over all identified waves is 87.7° from the local mean field. Scattering or spread from the perpendicular direction is relatively small (standard deviation about 12.0°) except for some deviations at higher wave numbers.

Figure 4.4 bottom panel displays the distribution of frequencies and wave numbers in the plasma rest frame. None of the three intervals exhibits a clear organization of dispersion relation; frequencies and wave numbers appear to be scattered. The spread in the frequency-wave number diagram can be interpreted as the sign that nonlinear energy cascade is operating. Interestingly, the distribution exhibits nevertheless a tendency that lower and higher frequencies are associated with smaller and larger wave numbers, respectively. Some waves are characterized by negative frequencies suggesting that solar wind turbulence contains both anti-sunward (radially outward from the Sun) and sunward (inward) propagating fluctuations in the plasma rest frame. Although the identified waves are spread in the frequency-wave

Fig. 4.4 Propagation angles from the mean magnetic field and the frequency-wave number diagram (dispersion diagram) of magnetic field fluctuations in the solar wind derived from Cluster data [32]

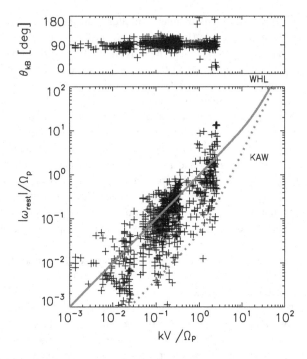

number diagram, there is a weaker result concerning their dense population. Two dispersion relations are over-plotted onto the diagram using the averaged plasma beta and propagation angles from the measurements: kinetic Alfvén waves (dotted line) and oblique whistler waves (solid line). Comparison yields a weak agreement with the oblique whistler mode, in the sense that the dispersion relation overlaps with the dense parts of wave populations in the frequency-wave number diagram.

It is interesting to note that other studies of dispersion relation using electric field data and the wave telescope/k-filtering suggest, on the other hand, that the kinetic Alfvén waves play a major role in solar wind turbulence [27, 29, 33]. This difference may be explained by a variety of solar wind conditions, e.g., high-speed and low-speed streams, plasma beta, and so on.

4.2 Turbulence Across the Bow Shock

4.2.1 Foreshock and Magnetosheath

Solar wind is a radially expanding, super-sonic and super-Alfvénic stream from corona. Shock waves form in various places in the heliosphere: as a standing shock (bow shock) in front of planetary magnetospheres and comets, and as a propagating, expanding shock wave accompanied by coronal mass ejection or by high-speed stream. Before spacecraft era, it was questioned if a shock wave really exists in space plasma, since the lack of particle collisions in the plasma might prevent from the shock formation. Nowadays, in contrast, shock waves can be found in many places in the heliosphere as mentioned above, and are still one of the major challenges in astrophysical plasma physics on the mechanism of energy dissipation in collisionless medium.

Earth's bow shock is one of the collisionless shocks, standing at about 15–20 Earth radii in front (on the dayside). It is the closest collisionless shock to us, and serves as a natural laboratory for studying dissipation mechanisms in the collisionless medium allowing for detailed, in-situ observations using spacecraft. Many spacecraft visited and encounterd the Earth bow shock, the interplanetary shocks, and bow shocks at other planets. Shock waves in plasma are accompanied by large-amplitude waves and turbulent fluctuations not only downstream but also upstream. Shock waves exist in astrophysical systems, too, and there are indications that interstellar shocks or supernova remnants are associated with turbulence [34, 35]. Shock waves play an important role in astrophysics, in particular acceleration of cosmic ray associated with supernova explosion.

One of the prominent differences between ordinary gas dynamic shock waves and collisionless shock waves is on the existence of foreshock. The mean velocity of the solar wind is typically about 400 km/s at Earth orbit (or 1 AU). Earth's magnetosphere is a blunt obstacle to the solar wind and the bow shock is formed as the solar wind encounters this obstacle. Spacecraft observations showed that collisionless shocks at sufficiently high Mach number have a unique dissipation mechanism, that is a portion of the incoming particles (electrons and ions) are accelerated and specularly reflected at the shock front. Various mechanisms have been proposed to account for acceleration. As a consequence, the reflected particles stream backward (upstream against the incoming flow) along the magnetic field. This forms an extended transition region in front of the shock in which the magnetic field lines are connected to the shock front. This transition region between the incoming high-Mach number flow and the shock front is called the foreshock. Because of the field line geometry, it is associated with the quasi-parallel shock (quasi-parallel between the magnetic field direction and the shock normal direction).

The foreshock region is characterized by large-amplitude fluctuations of magnetic field, flow velocity, and density. Shock-reflected, backstreaming ions form field-aligned beam along the magnetic field, and become unstable due to the presence of the solar wind ions (they form counter-streaming, two beam populations in the velocity space). The unstable velocity distribution collapses into a more stable, ring-shaped distribution around the magnetic field direction, and excites electromagnetic waves at low frequencies [36, 37]. A portion of electron population in the solar wind is also reflected at the shock. The backstreaming electrons excite high-frequency (kHz range), electrostatic broadband disturbance.

In contrast to the quasi-parallel shock, when the magnetic field is quasi-perpendicular to the shock normal direction, the transition into the downstream region takes place only on a small scale, of the order of several ion gyro-radii.

The bulk speed of the solar wind is diminished at the shock, and the flow kinetic energy is transformed into thermal energy (i.e., dissipation) and into electromagnetic energy (wave excitation). When the solar wind plasma enters the magnetosheath, downstream of the bow shock, the flow is deflected to pass by the magnetosphere. A part of the magnetosheath flow reaches the cusp region behind the dayside magnetosphere.

The magnetosheath is also characterized by large-amplitude fluctuations, in particular, of magnetic field and density. The reason for this is that thermalization of ions at the shock is an anisotropic process, preferring the directions perpendicular to the magnetic field due to their gyro-motion. The velocity distribution is therefore highly anisotropic; the thermal spread is large in the perpendicular direction and small in the parallel direction, and this is kinetically unstable, collapsing into a more stable distribution (becomes more isotropic) and exciting a spatially periodic, compressive structure in the magnetic field called the mirror mode. This is a special kind of wave, for the frequency is zero or so small that they do not propagate virtually (i.e., merely spatial structure). As they are swept by the flow, they are observed as a propagating wave due to Doppler shift.

The multi-point measurements of Cluster provided the wave vector spectra, the dispersion relations, and the spatial distribution of wave vectors or propagation directions for magnetic field fluctuations in the foreshock and the magnetosheath. The interaction between solar wind turbulence and the bow shock was studied by tracing the evolution of the wave vector spectra and the dispersion relations across the shock.

4.2.2 Energy Spectra and Anisotropy

Two Cluster shock-crossing events were investigated for studying turbulence evolution across the shock. The orbit of the first crossing (orbit A) encountered (1) the solar wind, (2) the foreshock, and (3) the magnetosheath; and the orbit for the second crossing (orbit B) encountered (4) the solar wind, (5) the magnetosheath, and (6) the magnetospheric cusp region. The first orbit represents a crossing of the quasi-parallel shock (Alfvén Mach number 7.6, angle between the shock normal and the upstream magnetic field 30.5°), the second one represents a crossing of the quasi-perpendicular shock (Alfvén Mach number 4.0, angle 75.7°).

4.2.2.1 Evolution in Frequency Domain

Figure 4.5 displays the frequency spectra for fluctuating magnetic field parallel and perpendicular to the mean field for these six regions. Upper panels are the spectra from Cluster crossing of the quasi-parallel shock: (1) solar wind, (2) foreshock, and (3) magnetosheath. Lower panels are the spectra from crossing of the quasi-perpendicular shock: (4) solar wind, (5) magnetosheath, and (6) cusp.

The frequency spectra in the solar wind confirm earlier observations using single-spacecraft: Incompressible fluctuations in the sense that the fluctuating fields are perpendicular to the mean field direction; power-law spectra that are reminiscent

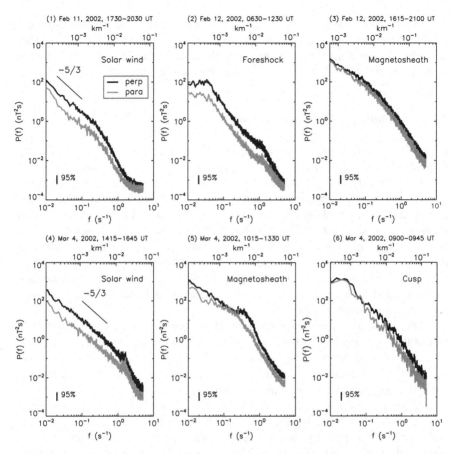

Fig. 4.5 Frequency spectra for magnetic field fluctuations parallel and perpendicular to the mean field in regions (1) solar wind, (2) foreshock, and (3) magnetosheath from Cluster crossing of the quasi-parallel shock; and (4) solar wind, (5) magnetosheath, and (6) cusp from crossing of the quasi-perpendicular shock. The scales on top of each panel are the wave numbers derived from spacecraft-frame frequencies using Taylor's hypothesis

of the inertial-range and the dissipation-range spectra for fluid turbulence. (Flatter curve at higher frequencies above 2 Hz represents the instrumental noise level)

In the foreshock, fluctuations perpendicular to the mean magnetic field dominate and the spectrum is characterized by two humps at lower frequencies around 0.04 Hz and at higher frequencies around 1 Hz. The former is interpreted as the pumping wave due to the backstreaming ions, but the physical process at the second hump is still unanswered.

In the magnetosheath, the spectral curves are different between that for the quasi-parallel shock and the quasi-perpendicular shock. Downstream of the quasi-parallel shock and in the cusp region, compressible fluctuations (parallel fluctuation with respect to the mean magnetic field) reach almost the spectral level for the incompressible fluctuations, and the fluctuation power is smoothly decaying toward higher frequencies. Downstream of the quasi-perpendicular shock, the spectrum of compressible fluctuations reach the level of incompressible fluctuations at intermediate frequencies around 0.1 Hz. Incompressible fluctuations exhibits a hump at higher frequencies around 0.4 Hz. The spectra exhibit different power-law regimes.

4.2.2.2 Evolution in Wave Vector Domain

Figure 4.6 displays the wave vector spectra for the same time intervals of Cluster displayed in Fig. 4.5. In the solar wind (region 1 and 4) the fluctuation energy is the smallest for the both shock crossings. In region 1 the distribution is extended in the perpendicular direction to the mean magnetic field and is elliptically shaped on an intermediate scale (about 0.010 rad/km), while it is moderately rectangular with dominant extension in the perpendicular direction at larger wave numbers (about 0.020 rad/km). In region 4 the distribution is also extended in the perpendicular direction. The ratio of the two reduced spectra exhibits a negative trend down to values $0.8 - 0.9$ toward larger wave numbers in the both cases, reflecting the dominance of the perpendicular wave vector geometry. These results justify the picture of two-dimensional turbulence in the solar wind [13, 14].

In the foreshock region (region 2) the fluctuation energy becomes larger than that of the solar wind by factor about 10. In contrast to the solar wind, the energy distribution is extended in the parallel direction to the mean magnetic field and represents the dominance of the parallel wave vector geometry, which is reflected in the anisotropy ratio as a positive trend up to the ratio 1.5.

In the magnetosheath (region 3 and 5) the fluctuation energy is further enhanced from the foreshock by factor about 10 (region 3) and from the solar wind by factor about 50 (region 5). The distribution in region 3 exhibits an asymmetric feature between the parallel and the anti-parallel directions as well as anisotropy between the parallel and the perpendicular directions. The ratio of the two reduced spectra exhibits a negative trend down to 0.6 and suggests the dominance of the perpendicular wave vector geometry. Anisotropy preferring the perpendicular wave vector geometry is stronger than that of the solar wind (region 1). In region 5 the fluctuation energy is enhanced from the solar wind across the shock while the distribution maintains the extended structure in the perpendicular direction. Also, there is a moderate asymmetry between the parallel and the anti-parallel directions. The ratio of the reduced spectra exhibits a negative trend, preferring the perpendicular wave vector geometry, and the anisotropy ratio curve is very similar to that of the solar wind (region 4).

In the cusp region (region 6) the fluctuation energy is enhanced from the magnetosheath and the energy distribution is further extended in the perpendicular direction. The anisotropy ratio curve exhibits a negative trend 0.5, which may suggest that cusp turbulence inherits the properties of the magnetosheath fluctuations.

To summarize, there is a clear transition along the trajectory of the quasi-parallel shock crossing: perpendicular wave vector geometry in the solar wind, parallel geometry in the foreshock, and back to perpendicular geometry in the magnetosheath. In

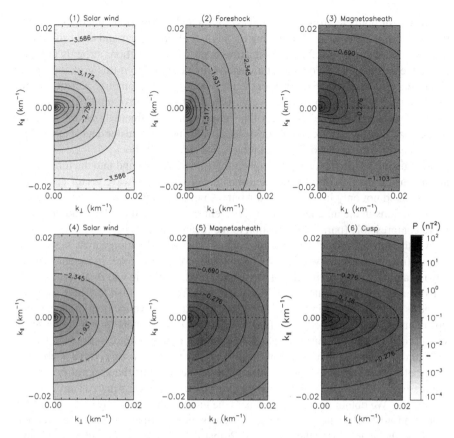

Fig. 4.6 Energy spectra in the wave vector domain for six regions across the bow shock displayed in Fig. 4.5. The spectra are derived from Cluster data, and are averaged over the directions around the mean magnetic field [38]

contrast, along the trajectory of the quasi-perpendicular shock crossing the wave vector spectra exhibit persistently perpendicular wave vector geometry.

Likely sources for the anisotropies of turbulent fluctuations are wave-wave interactions and plasma instabilities. In the foreshock, ion-beam modes are the most likely instabilities because the bow shock is a source of heated electrons and reflected ions. In the frequency range of concern to us (up to about ion gyro-frequency), the most likely mode to grow is the electromagnetic ion/ion right-hand resonant instability. This has been confirmed observationally by Watanabe and Terasawa [39] and Fuselier et al. [40] using single spacecraft methods and recently by multi-spacecraft methods of Cluster [41, 42]. Linear kinetic theory [43] shows that the instability has maximum growth in the parallel and anti-parallel directions to the mean magnetic field, so that enhanced fluctuations from this instability should have properties of the parallel wave vector geometry. Magnetosheath plasma shows the consequences of magnetic compression and heating at the shock, so that the primary characteristic of proton distributions in this regime is a strong temperature anisotropy. This anisotropy leads to the growth of both electromagnetic ion cyclotron and mirror mode fluctuations with the mirror instability often dominating the high-β plasmas near and downstream of the shock. The mirror instability has maximum growth at directions strongly oblique to the mean magnetic field, so that enhanced fluctuations from this growing mode should have properties of the perpendicular wave vector geometry. Anisotropy in the cusp region can also be qualitatively explained by the mirror mode fluctuations. Luhmann et al. [44] argued that one of the major sources of fluctuations downstream of the quasi-parallel shock is the foreshock activity, but the Cluster data suggest that the property of the foreshock fluctuations is lost across the shock.

Fluctuation properties are lost at transition from the solar wind into the foreshock, and again at transition into the magnetosheath. Perpendicular wave vector geometry can be found not only in the solar wind but also in the magnetosheath, but the sense of fluctuations in the magnetosheath is different from that in the solar wind: they are more compressible (parallel fluctuation to the mean field).

Another process which may be relevant is wave amplification across the shock. Fluctuations in the solar wind may be amplified at the bow shock independently of any instabilities. The interaction of the magnetohydrodynamic waves with the shock wave was analytically studied by McKenzie and Westphal [45–47]. Their analysis suggests that the magnetic field amplitude of an Alfvén wave incident in the shock-upstream region is enhanced by a factor of unity or three, depending on the sense of wave propagation in the upstream and downstream region with respect to the shock normal direction and that the amplification of a fast magnetosonic wave is about a factor of four. Therefore a naïve estimate gives the jump of the spectral power by factors $10-20$ across the shock, as energy is proportional to the squared amplitude of fluctuation. We obtain in our measurement the jump of the energy by factor about 10 from the foreshock to the magnetosheath across the quasi-parallel shock, and about 50 across the quasi-perpendicular shock. Probably there are a variety of mechanisms that contribute to the amplification across the shock such as wave mode conversion and wave reflection at the shock or at the magnetopause. In this analysis it is difficult to distinguish between plasma instabilities and shock amplification effects because only the total energy is used in the analysis. Performing polarization or helicity analysis is also possible, and this will help to distinguish these two effects, as such an analysis can determine energy spectra for different fluctuation components and for different wave vectors.

4.2.3 Dispersion Relation

Transitions in wave or turbulence properties across the bow shock can also be found in the dispersion relations (or frequency-wave number diagrams). Figure 4.7 displays two examples of dispersion relations obtained from Cluster data: in the foreshock (left panel) and in the magnetosheath (downstream of the quasi-parallel shock, right panel). In contrast to solar wind turbulence, fluctuations in the foreshock and the magnetosheath are well characterized by normal modes and the concept of dispersion relation is valid.

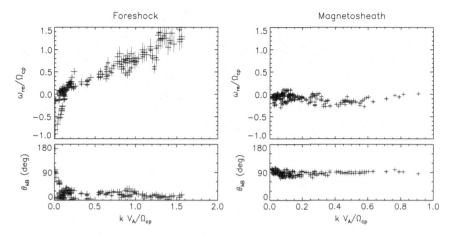

Fig. 4.7 Frequency-wave number diagrams in the foreshock and in the magnetosheath (top panels) derived from Cluster data. Frequencies are corrected for the Doppler shift (into the flow rest frame or co-moving frame) and normalized to the proton cyclotron frequency. Wave numbers are normalized to that for the proton inertial length. Lower panels are the angles of the wave vectors from the mean magnetic field direction [42]

In the foreshock region, the frequency-wave number diagram exhibits two distinct branches. One is nearly parallel propagating Alfvén wave branch (with respect to the mean magnetic field direction), and the other is the beam-resonant mode in which wave speeds (group speeds) are the same as the backstreaming ion-beam speed. The former is the dominant mode and has larger amplitudes. Dispersion analysis of another time interval using Cluster data shows that the dominant branch represents the right-hand mode that continues to the whistler wave at higher frequencies [41].

In the magnetosheath, the rest-frame frequencies are small, well below the proton cyclotron frequencies in a broad range of wave numbers up to the wave number for the proton inertial length. Propagation angle is nearly perpendicular to the mean magnetic field. This branch can be explained by the mirror mode associated with temperature anisotropy caused by shock dissipation.

4.2.4 Propagation Pattern

Cluster observes fluctuations of magnetic field at various positions in the foreshock and the magnetosheath, from the bow shock to the magnetopause at various zenith angles, and at various distances from the Sun-Earth line. These samples of waves can be reduced into a two-dimensional plot, as spatial distribution of propagation speeds and directions averaged over the directions around the Sun-Earth axis and over segments along the shape of the bow shock and the magnetopause (Fig. 4.8).

The foreshock waves propagate mostly toward upstream away from the bow shock (in the co-moving frame). The sense of propagation is the same as that of backstreaming ions from the shock. In contrast, the sense of propagation in the magnetosheath is

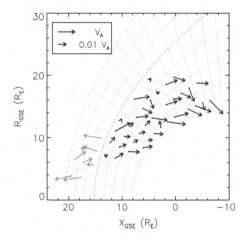

Fig. 4.8 Spatial distribution of wave propagation speeds and directions in the flow rest frame (co-moving frame) averaged over segments along the shape of the magnetopause and the bow shock, derived from Cluster data. The phase speeds are normalized to the Alfvén speed and plotted as arrows on a logarithmic scale. The arrows in gray represent the foreshock waves, and the arrows in black the magnetosheath waves. The dotted curve is the nominal position of the bow shock [48]

pointing oppositely from the foreshock waves. They propagate further downstream away from the shock (again in the co-moving frame). It is true that the mirror mode is in the simplest picture a zero-frequency mode and thus does not propagate, but Cluster measurements also presented that the rest-frame frequencies and phase speeds are still finite in the magnetosheath, and furthermore organize a large-scale organization or propagation pattern, which can also be seen in the directions around the Sun-Earth axis [49]. The anti-sunward propagation dominates in the magnetosheath. At the smaller zenith angles the propagation is toward the magnetosheath flank, and at the larger angles it is toward the magnetopause.

An interesting relationship was also found observationally between the propagation pattern and the density distribution in the magnetosheath [49]. The propagation directions are almost perpendicular to the contours of density distribution, i.e., along density gradients. Waves may be coupled to the density gradient, becoming the so-called drift waves. The drift mirror mode, however, propagates perpendicular to the density gradient direction [50, 51], whereas Cluster data analysis showed the parallel direction. Perhaps the drift mirror mode is operating in the magnetosheath, but the frequency is proportional to logarithm of the density distribution, $\omega \propto \nabla \log n$, and therefore the wave drift effect would be too small to detect. The parallel propagation to the density gradient may suggest a wave refraction in an inhomogeneous medium. In that case the WKB approximation or the ray tracing method should be applied to the low frequency waves in the magnetosheath, which solves the wave vector $\mathbf{k}$ as function of the coordinate $\mathbf{r}$.

Table 4.1 Fluctuation properties in the solar wind, the foreshock, and the magnetosheath

Region	Solar wind	Foreshock	Magnetosheath
Fluctuation sense	Incompressible	Incompressible	Compressible
Wave vector geometry	Perpendicular	Parallel	Perpendicular
Dispersion relation	Unclear	Clear	Clear
Wave modes	Oblique whistler	Right-hand resonant	Mirror mode
	Kinetic Alfvén	Beam resonant	

4.3 Summary

Plasma turbulence essentially differs from fluid turbulence in many aspects. There are a great number of normal modes and non-normal modes that can serve as the carrier of energy cascade through wave–wave scattering; Energy dissipation cannot be made by particle collisions and must be by wave-particle interactions; The presence of large-scale magnetic field imposes anisotropy and asymmetry in the energy spectra; Fluctuation sense can also be anisotropic (compressible and incompressible) with respect to the large-scale field.

Cluster observations revealed that fluctuation properties are considerably different from one region to another in near-Earth space. These properties are summarized in Table 4.1. Fluctuation sense can be compressible (magnetosheath) and incompressible (solar wind and foreshock); Wave vector geometry of the energy spectra can be parallel (foreshock) and perpendicular (solar wind and magnetosheath); The existence of dispersion relation was confirmed for the foreshock and the magnetosheath, while solar wind turbulence seems to be a mixture of both normal and non-normal modes.

It is interesting to note that fluctuation amplitude becomes enhanced from the solar wind to the foreshock, and further enhanced to the magnetosheath. However, there is only weak evidence of dispersion relation in the solar wind, whereas the foreshock and the magnetosheath exhibits clear dispersion relations. This suggests that nonlinear processes generating non-normal modes can be significant even fluctuation amplitudes are small in plasma, and that turbulence must be a widespread phenomenon in astrophysical systems as turbulence needs only small amplitudes for its generation.

Because of the assumption of plane wave geometry used in the analysis, one cannot rule out the possibility of other fluctuation geometries such as spherical wave patterns or phase-shifted waves. Turbulent fluctuations in the solar wind may also have various origins, including coronal magnetic field structures and solar wind expansion effect. How much energy is transported by fluctuations outward and inward with respect to the sun on the coronal level would be a valuable material for further discussion.

References

1. Coleman, P.J. Jr: Turbulence, viscosity, and dissipation in the solar-wind plasma. Astrophys. J. **153**, 371–388 (1968)
2. Matthaeus, W.H., Goldstein, M.L.: Measurement of the rugged invariants of magnetohydrodynamic turbulence in the solar wind. J. Geophys. Res. **87**, 6011–6028 (1982)
3. Matthaeus, W.H., Goldstein, M.L., Smith, C.: Evaluation of magnetic helicity in homogeneous turbulence. Phys. Rev. Lett. **48**, 1256–1259 (1982)
4. Marsch, E., Tu, C.-Y.: On the radial evolution of MHD turbulence in the inner heliosphere. J. Geophys. Res. **95**, 8211–8229 (1990)
5. Leamon, R.J., Smith, C.W., Ness, N.F., Matthaeus, W. H.: Observational constraints on the dynamics of the interplanetary magnetic field dissipation range. J. Geophys. Res. **103**, 4775–4787 (1998)
6. Podesta, J.J., Roberts, D.A., Goldstein, M.L.: Spectral exponents of kinetic and magnetic energy spectra in solar wind turbulence. Astrophys. J. **664**, 543–548 (2007)
7. Marsch, E., Tu, C.-Y.: Intermittency, non-Gaussian statistics and fractal scaling of MHD fluctuations in the solar wind. Nonlin. Processes Geophys. **4**, 101–124 (1997)
8. Marsch, E., Goertz, C.K., Richter, K.: Wave heating and acceleration of solar wind ions by cyclotron resonance. J. Geophys. Res. **87**, 5030 (1982)
9. Marsch, E., Tu, C.-Y.: Evidence for pitch angle diffusion of solar wind protons in resonance with cyclotron waves. J. Geophys. Res. **106**, 8357–8362 (2001)
10. Marsch, E., Tu, C.-Y.: Anisotropy regulation and plateau formation through pitch angle diffusion of solar wind protons in resonance with cyclotron waves. J. Geophys. Res. **107**, 1249 (2002)
11. Bieber, J.W., Matthaeus, W.H., Smith, C.W., Wanner, W., Kallenrode, M.-B., Wibberenz, G.: Proton and electron mean free paths: The Palmer consensus revisited. Astrophys. J. **420**, 294–306 (1994)
12. Bieber, J.W., Wannger, W., Matthaeus, W.H.: Dominant two-dimensional solar wind turbulence with implications for cosmic ray transport. J. Geophys. Res. **101**, 2511–2522 (1996)
13. Matthaeus, W. H., Goldstein, M. L., Roberts, D. A.: Evidence for the presence of quasi-two-dimensional nearly incompressible fluctuations in the solar wind, J. Geophys. Res. **95**, 20673–20683 (1990)
14. Carbone, V., Malara, F., Veltri, P.: A model for the three-dimensional magnetic field correlation spectra of low-frequency solar wind fluctuations during Alfvénic periods. J. Geophys. Res. **100**, 1763–1778 (1995)
15. Dasso, S., Milano, L.J., Matthaeus, W.H., Smith, C.W.: Anisotropy in fast and slow solar wind fluctuations. Astrophys. J. **635**, L181–L184 (2005)
16. Pinçon, J.L., Lefeuvre, F.: Local characterization of homogeneous turbulence in a space plasma from simultaneous measurement of field components at several points in space. J. Geophys. Res. **96**, 1789–1802 (1991)
17. Motschmann, U., Woodward, T.I., Glassmeier, K.-H., Southwood, D.J., Pinçon, J.L.: Wavelength and direction filtering by magnetic measurements at satellite arrays: generalized minimum variance analysis. J. Geophys. Res. **101**, 4961–4965 (1996)
18. Pinçon, J.-L., Motschmann, U.: Multi-Spacecraft Filtering: General Framework, Analysis Methods for Multi-Spacecraft Data. In: Paschmann G., Daly P.W., (eds.) ISSI Sci. Rep. SR-001, chap. 3. International Space Science Institute, Berne, Switzerland, pp. 65–78 (1998)
19. Glassmeier, K.-H., Motschmann, U., Dunlop, M., Balogh, A., Acuña, M.H., Carr, C., Musmann, G., Fornaçon, K.-H., Schweda, K., Vogt, J., Georgescu, E., Buchert, S.: Cluster as a wave telescope - first results from the fluxgate magnetometer, Ann. Geophys. 19, 1439–1447 (2001) Correction in 21, 1071 (2003).
20. Rème, H., Aoustin, C., Bosqued, J.M., Dandouras, I., Lavraud, B., Sauvaud, J.A., Barthe, A., Bouyssou., , Camus T, h., Coeur-Joly, O., Cros, A., Cuvilo, J., Ducay, F., Garbarowitz, Y., Medale, J.L., Penou, E., Perrier, H., Romefort, D., Rouzaud, J., Vallat, C., Alcaydé, D., Jacquey,

C., Mazelle, C., D'Uston, C., Möbius, E., Kistler, L.M., Crocker, K., Granoff, M., Mouikis, C., Popecki, M., Vosbury, M., Klecker, B., Hovestadt, D., Kucharek, H., Kuenneth, E., Paschmann, G., Scholer, M., Sckopke, N., Seidenschwang, E., Carlson, C.W., Curtis, D.W., Ingraham, C., Lin, R.P., McFadden, J.P., Parks, G.K., Phan, T., Formisano, V., Amata, E., Bavassano-Cattaneo, M.B., Baldetti, P., Bruno, R., Chionchio, G., di Lellis, A., Marcucci, M.F., Pallocchia, G., Korth, A., Daly, P.W., Graeve, B., Rosenbauer, H., Vasyliunas, V., McCarthy, M., Wilber, M., Eliasson, L., Lundin, R., Olsen, S., Shelley, E.G., Fuselier, S., Ghielmetti, A.G., Lennartsson, W., Escoubet, C.P., Balsiger, H., Friedel, R., Cao, J.-B., Kovrazhkin, R.A., Papamastorakis, I., Pellat, R., Scudder, J., Sonnerup, B.: First multispacecraft ion measurements in and near the Earth's magnetosphere with the identical Cluster ion spectrometry (CIS) experiment. Ann. Geophys. **19**, 1303–1354 (2001)

21. Narita, Y., Glassmeier, K.-H., Sahraoui, F., Goldstein, M.L.: Wave-vector dependence of magnetic-turbulence spectra in the solar wind. Phys. Rev. Lett. **104**, 171101 (2010) Doi: 10.1103/PhyRevLett104.171101

22. Narita, Y., Glassmeier, K.-H., Goldstein, M. L., Motschmann, U., and Sahraoui, F.: Three-dimensional spatial structures of solar wind turbulence from 10,000-km to 100-km scales, Ann. Geophys. **29**, 1731-1738 (2011)

23. Podesta, J.J.: Dependence of solar-wind power spectra on the direction of the local mean magnetic field. Astrophys. J. **698**, 986–999 (2009) Doi:1.1088/0004-637X/698/2/986

24. Parashar, T.N., Shay, M.A., Cassak, P.A., Matthaeus, W.H.: Kinetic dissipation and anisotropic heating in a turbulent collisionless plasma. Phys. Plasmas. **16**, 032310 (2009)

25. Svidzinski, V.A., Li, H., Rose, H.A., Albright, B.J., Bowers, K.J.: Particle in cell simulations of fast magnetosonic wave turbulenc in the ion cyclotron frequency range. Phys. Plasmas **16**, 122310 (2009)

26. Dmitruk, P., Matthaeus, W.H.: Waves and turbulence in magnetohydrodynamic direct numerical simulations. Phys. Plasmas **16**, 062304 (2009)

27. Sahraoui, F., Goldstein, M.L., Robert, P., Khotyaintsev, Yu.V.: Evidence of a cascade and dissipation of solar-wind turbulence at the electron gyroscale. Phys. Rev. Lett. **102**, 231102 (2009)

28. Perri, S., Carbone, V., Veltri, P.: Where does fluid-like turbulence break down in the solar wind?. Astrophys. J. Lett. **725**, L52–L55 (2010)

29. Bale, S.D., Kellogg, P.J., Mozer, F.S., Horbury, T.S., Reme, H.: Measurement of the electric fluctuation spectrum of magnetohydrodynamic turbulence. Phys. Rev. Lett. **94**, 215002 (2005) Doi:10.1103/PhysRevLett.94.215002

30. Matthaeus, W.H., Servidio, S., Dmitruk, P.: Comment on "Kinetic simulations of magnetized turbulence in astrophysical plasmas". Phys. Rev. Lett. **101**, 149501 (2008)

31. Narita, Y., Glassmeier, K.-H., Motschmann, U.: Wave vector analysis methods using multi-point measurements. Nonlin. Processes Geophys. **17**, 383–394 (2010)

32. Narita, Y., Gary, S.P., Saito, S., Glassmeier, K.-H., Motschmann, U.: Dispersion relation analysis of solar wind turbulence. Geophys. Res. Lett. **38**, L05101 (2011) Doi:10.1029/2010GL046588

33. Sahraoui, F., Goldstein, M.L., Belmont, G., Canu, P., Rezeau, L.: Three dimensional anisotropic k spectra of turbulence at subproton scales in the solar wind. Phys. Rev. Lett. **105**, 131101 (2010) Doi:10.1103/PhysRevLett.105.131101

34. Hester, J.J., Raymond, J.C., Blair, W.P.: . Astrophys. J. **420**, 721–745 (1994)

35. Spitler, L.G., Spangler, S.R.: Limits on enhanced radio wave scattering by supernova remnants. Astrophys. J. **632**, 932–940 (2005)

36. Paschmann, G., Sckopke, N. Bame S.J., Asbridge, J.R., Gosling, J.T., Russell, C.T., Greenstadt, E.W.: Association of low-frequency waves with suprathermal ions in the upstream solar wind. Geophys. Res. Lett. **6**, 209–212 (1979)

37. Paschmann, G., Sckopke, N., Papamastorakis, I., Asbridge, J.R., Bame, S.J., Gosling, J.T.: Characteristics of reflected and diffuse ions upstream from the Earth's bow shock. J. Geophys. Res. **86**, 4355–4364 (1981)

38. Narita, Y., and Glassmeier, K.-H.: Anisotropy evolution of magnetic field fluctuation through the bow shock, Earth Planets Space, 62, e1–e4, (2010)
39. Watanabe, Y., Terasawa, T.: On the excitation mechanism of the low-frequency upstream waves. J. Geophys. Res. **89**, 6623–6630 (1984)
40. Fuselier, S.A., Thomsen, M.F., Gosling, J.T., Bame, S.J., Russell, C.T.: Gyrating and intermediate ion distributions upstream from the earth's bow shock. J. Geophys. Res. **91**, 91–99 (1986)
41. Narita, Y., Glassmeier, K.-H., Schäfer, S., Motschmann, U., Sauer, K., Dandouras, I., Fornaçon, K.-H., Georgescu, E., Rème, H.: Dispersion analysis of ULF waves in the foreshock using cluster data and the wave telescope technique. Geophys. Res. Lett. **30**, SSC 43–1 (2003)
42. Narita, Y., Glassmeier, K.-H.: Dispersion analysis of low-frequency waves through the terrestrial bow shock. J. Geophys. Res. **110**, A12215 (2005) Doi:10.1029/2005JA011256
43. Gary, S. P.: Theory of space plasma microinstabilities. Cambridge Atmospheric and Space Science Series, Cambridge (1993)
44. Luhmann, J.G., Russell, C.T., Elphic, R.C.: Spatial distributions of magnetic field fluctuations in the dayside magnetosheath. J. Geophys. Res. **91**, 1711–1715 (1986)
45. McKenzie, J.F., Westphal, K.O.: Transmission of Alfvén waves through the Earth's bow shock. Planet. Space Sci. **17**, 1029–1037 (1969)
46. McKenzie, J.F., Westphal, K.O.: Interaction of hydromagnetic waves with hydromagnetic shocks. Phys. Fluids. **13**, 630–640 (1970)
47. McKenzie, J.F.: Hydromagnetic wave interaction with the magnetopause and the bow shock. Planet. Space Sci. **18**, 1–23 (1970)
48. Narita, Y., Glassmeier, K.-H., Fornaçon, K.-H., Richter, I., Schäfer, S., Motschmann, U., Dandouras, I., Rème, H., Georgescu, E.: Low frequency wave characteristics in the upstream and downstream regime of the terrestrial bow shock. J. Geophys. Res. **111**, A01203 (2006) Doi:10.1029/2005JA011231
49. Narita, Y., Glassmeier, K.-H.: Propagation pattern of low-frequency waves in the terrestrial magnetosheath. Ann. Geophys. **24**, 2441–2444 (2006)
50. Hasegawa, A.: Drift mirror instability in the magnetosphere. Ann. Fluids. **12**, 2642 (1969)
51. Pokhotelov, O.A., Balikhin, M.A. Treumann R.A., Pavlenko, V.P.: Drift mirror instability revisited: 1. cold electron temperature limit. J. Geophys. Res. **106**, 8455–8464 (2001)

Chapter 5
Impacts on Related Subjects

Abstract Plasma turbulence is not only a challenge in physics and mathematics for its nonlinear complexities, but also has impacts on astrophysics, solar system science, and Earth science. This chapter discusses the physical picture of plasma turbulence derived from multi-point measurements of Cluster and its implications to these related subjects.

5.1 Plasma Physics and Turbulence Theories

5.1.1 Wave–Wave Interactions

Fluctuations of the magnetic field have different characteristics, in particular between the solar wind and the foreshock. The former is characterized by smaller amplitudes, perpendicular wave vectors (to the large-scale magnetic field), and a weak signature of dispersion relation; and the latter by larger amplitudes, parallel wave vectors, and a clear dispersion relation. This raises a question why the foreshock fluctuations display a clear dispersion relation in spite of larger amplitudes. The theoretical concept of dispersion relation is based on linearization of dynamical equation (MHD equation, for example) under which fluctuation amplitudes are assumed to be so small compared to the mean field. In reality it turned out to be opposite: small-amplitude fluctuations do not show clear dispersion curve, but large-amplitude fluctuations do.

Different scenarios are possible to explain this contradictory situation between the solar wind and the foreshock. The first possibility is that fluctuations in these two regions have completely different evolution scenarios. The second possibility is that solar wind might have had a dispersion relation but such information was lost during the travel from the solar corona to the interplanetary space.

In the first scenario, the foreshock fluctuations represent "waves" characterized by a dispersion relation, whereas the solar wind fluctuations represent fully developed turbulence. In this interpretation, wave–wave interactions (or energy cascade) are

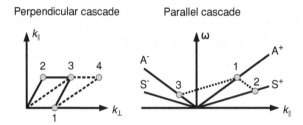

Fig. 5.1 Wave–wave interactions causing energy transport or cascade in different directions of the mean magnetic field. *Left* panel is the Alfvén wave interactions generating increasingly oblique wave vectors. *Right* panel is the wave interactions for decay instability in which an Alfvén wave is split into a forward sound wave and a backward Alfvén wave with respect to the direction of the parent wave

operating in different fashions. Solar wind turbulence may evolve as a consequence of perpendicular energy cascade. Suppose three-wave interaction is operating for Alfvén waves. Wave resonance condition holds for frequencies and wave vectors:

$$\omega_1 + \omega_2 = \omega_3 \tag{5.1}$$

$$\mathbf{k}_1 + \mathbf{k}_2 = \mathbf{k}_3. \tag{5.2}$$

The frequency condition can be written into a wave number condition for Alfvén wave dispersion relation, $\omega = \pm k_\parallel v_\text{A}$,

$$k_{\parallel 1} - k_{\parallel 2} = \pm k_{\parallel 3}. \tag{5.3}$$

Hence the equations yield that $k_{\parallel 1} = 0$ or $k_{\parallel 2} = 0$ [1]. The left panel of Fig. 5.1 displays the wave evolution in the $k_\parallel - k_\perp$ diagram. The wave 1 is perpendicular to the mean magnetic field and it represents a non-propagating, spatial structure (cf. dispersion relation, $\omega = k_\parallel v_\text{A}$). It interacts with the wave 2 which has the wave vector in an oblique direction. The wave resonance condition implies that a new mode appears at the point completing the parallelogram with the wave 1 and the wave 2 in the $k_\parallel - k_\perp$ diagram. This new mode (the wave 3) is now more oblique than the wave 2 and can generate another mode (the wave 4) with the wave vector in even more oblique direction when it interacts recursively with the wave 1. This presents a picture that turbulence evolves in the sense of enhancing anisotropy preferring perpendicular wave vectors. This picture, however, still assumes that the Alfvén wave dispersion relation exists.

Energy cascade or the picture of wave–wave interactions can also be established in the parallel direction to the mean magnetic field, which might explain parallel wave vectors in the foreshock. The right panel of Fig. 5.1 displays the wave–wave interaction in the frequency-wave number diagram in the direction of the mean magnetic

Fig. 5.2 Scenarios of energy transport in the frequency-wave number diagram for fluid turbulence (no dispersion relation) and for plasma turbulence (with dispersion relation)

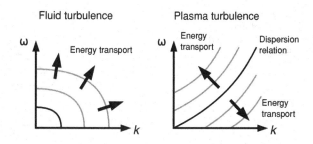

field. There are four different wave branches (dispersion relations): two of them are the Alfvén mode parallel and anti-parallel to the mean magnetic field (A^+ and A^-) and the other two are the sound mode along the magnetic field (S^+ and S^-). In this diagram it is assumed that the Alfvén speed v_A is larger than the sound speed. In the initial state only one wave mode exists (the wave 1) on the A^+ branch, and this breaks up into two different modes, A^- and S^+, while keeping the wave resonance condition for frequencies and wave numbers (or the parallelogram in the diagram). This process is referred to as the decay instability, suggesting that a large-amplitude Alfvén wave ends up with a number of Alfvén waves and sound waves when the instability or the wave splitting proceeds successively. Parallel energy cascade can also be constructed solely in the forward direction, too, without exciting daughter waves in the opposite direction (modulational instability). The wave–wave interactions of decay and modulation instabilities are also referred to as the parametric instability [2–4].

The second possibility of turbulence evolution focuses more on the elapsed time after the wave excitation. One of the difference between fluid turbulence and plasma turbulence is the existence of dispersion relation. The former conceptually represents the Richardson-Kolmogorov picture of eddies splitting into smaller sizes. Richardson found the scaling law between the separation of fluid elements (in the Lagrangian picture of fluid) and the elapsed time, and that is $|\Delta r|^2 \propto t^3$ [5]. Naïvely speaking, this scaling law suggests that energy cascade in fluid turbulence occurs under the constraint $\omega \propto k^{2/3}$ (Fig. 5.2. left panel). Plasmas, in contrast, allow different kinds of dispersion relations to exist and fluctuations may evolve from dispersion relation (i.e., normal mode waves) into fully developed turbulence deviating from the dispersion curve in the frequency-wave number diagram (Fig. 5.3. right panel). Thus, it is important to study the role of dispersion relation in plasma turbulence.

5.1.2 Wave-Particle Interactions

Frequency spectra of magnetic field fluctuation often exhibit a spectral break, and the spectral curve becomes steeper on smaller scales or at higher frequencies. This is reminiscent of the dissipation-range spectrum in Kolmogorov's phenomenological model, but on the other hand recent Cluster observations have demonstrated that the

Fig. 5.3 Fluctuation
geometry of magnetic flux
tube in the solar wind
turbulence

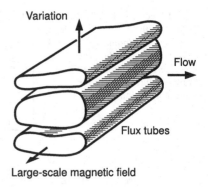

frequency spectra for electric field becomes flatter at higher frequencies, suggesting the concept of dispersion range is valid [6]. That is, energy is not converted into thermal energy (i.e., dissipation) but converted into electric field energy because waves are dispersive in that range. The phase speed is proportional to the ratio of electric to magnetic field amplitude, $v_{ph} \propto |\delta E|/|\delta B|$ and the dispersion range or a flatter spectral curve of the electric field implies that the phase speeds increases at higher frequencies.

Various kinds of wave-particle interactions are, however, possible in plasma turbulence, and they certainly play a role of energy dissipation. The interactions with charged particles are made primarily via wave electric field. Two mechanisms are particularly relevant: cyclotron resonance and Landau resonance. The former is resonance with rotating, perpendicular electric field with respect to the mean magnetic field direction, and the latter is resonance with parallel electric field, accelerating ions and electrons along the mean magnetic field. Dissipation might co-exist in the dispersion range. In that case, energy is transported onto smaller scale by dispersive wave modes and at the same time the energy is converted into thermal energy via wave-particle interactions. Again, it is important to study if fully-developed turbulence in plasma still contains dispersion relation, and if any, what kind of wave modes, e.g., kinetic Alfén waves, whistler waves, or other modes, are responsible for the wave-particle interaction.

5.1.3 Turbulence Interaction with Shock Waves

Turbulence fluctuations also appear downstream of the shock wave. One of the sources of turbulent fluctuations is solar wind turbulence and foreshock waves transmitted and processed at the shock wave. The linear analysis (or perturbative study) of Rankine-Hugoniot relation stating the conservation law for mass, momentum, energy, and electromagnetic field across the discontinuity suggests that mode conversion occurs between the three MHD wave modes: Alfvén, fast, and slow mode [7–9]. However, the use of MHD waves is questionable. Magnetosheath fluctuations exhibit the properties of the mirror mode which is a kinetic process associated with compressible fluctuations (parallel to the large-scale field direction), and very small frequencies in the co-moving frame such that the fluctuations virtually do not propagate but represent structures swept by the flow. The mirror mode is caused by the instability related with temperature anisotropy caused by the shock dissipation mechanism that ions are accelerated perpendicular to the mean magnetic field gyrate around that direction [10].

5.2 Solar System Science and Astrophysics

5.2.1 Lessons From Solar Wind Turbulence

Cluster data analysis visualized the three-dimensional spectrum of solar wind turbulence: it is extended in the direction perpendicular to the large-scale magnetic field and also perpendicular to the flow. This means that the fluctuation geometry has a variation (i.e., various wavelengths) in that direction in the coordinate space. Figure 5.3 displays the illustration of the fluctuation geometry derived from the Cluster data analysis. It represents the geometry for two-dimensional turbulence: the large-scale magnetic field behaves like a solid rod and it is hard to bend it but easy to displace field lines from one position to another (without bending the field lines). Extended Cluster data analysis furthermore identified that this property can be found on various scales in solar wind turbulence from 10,000 km down to 100 km [11].

The breakdown of axisymmetry around the large-scale magnetic field must represent some mechanisms in solar wind turbulence. One of likely mechanisms is the effect of divergent, radial expansion of solar wind from corona to the interplanetary space. The fluctuation geometry in the coordinate space (Fig. 5.3) suggests that variation or energy cascade is suppressed in the direction of flow, which could be interpreted as the picture that the fluctuating fields (or flux tubes) are stretched in that direction. Energy cascade process is constrained to the perpendicular direction, and it is the north-south component (with respect to the rotation of the Sun) in the solar equatorial plane (or the orbital planes of the planets). This picture of fluctuation geometry (one-dimensional variation) might be applicable to other astrophysical systems with radial expanding flow: stellar wind and galactic wind.

5.2.2 Turbulence in Other Astrophysical Systems

Plasma turbulence can be found in various phenomena or systems in astrophysics. Most known examples are acceleration and transport of cosmic ray, interstellar medium, and accretion disks.

Cosmic Ray

Cosmic ray (or galactic cosmic ray) is a homogeneous background of high-energy particles in a very broad energy range from 10^6 eV to 10^{20} eV. Majority of particle population is protons, and minor species are electrons, helium nuclei, and heavier elements. One of important effects of cosmic rays is to maintain ionization of the gas in the interstellar space, although its contribution to ionization is rather low. Cosmic ray has several interesting properties: (1) it is almost isotropically distributed when observed at the Earth; (2) its energy spectrum displays a power-law curve with the exponent between -2 and -3 [12]. Turbulence is believed to play an important role to explain these properties, in particular its acceleration mechanism and transport in the galaxy after acceleration.

The fluctuation geometry obtained from the Cluster data analysis in the solar wind will be an important ingredient for constructing a realistic model of cosmic ray processes. The notion of cosmic ray acceleration by shock waves was first discussed by Fermi, in particular by expanding shock waves caused by supernova explosion. When plasma turbulence exists both upstream and downstream of the shock wave and when the turbulent fluctuation scatters charged particles selectively from upstream to downstream and vice versa, some particles can be trapped by turbulence on the both sides of the shock wave. Since the upstream flow speed is higher than that of downstream, these particles gain energy in a statistical sense as they feel the convergent flow due to larger flow speed in the upstream region. This process is referred to as the diffusive shock acceleration or the second-order Fermi acceleration [13], cf., the first-order Fermi acceleration describes particles acceleration by trapping between two or more moving shock waves.

Also, fluctuation geometry of turbulence relative to the large-scale magnetic field is important for transport of cosmic rays: they travel along magnetic field lines and experience random oscillations on its trajectory. If the fluctuation geometry is more like a slab geometry associated with wave vectors parallel to the large-scale magnetic field or more like a two-dimensional turbulence geometry such that both fluctuations and wave vectors are confined in the plane perpendicular to the large-scale field is one of the critical questions in cosmic ray transport in the galaxy. Bieber et al. [14, 15] found that the two-dimensional turbulence geometry must play a dominant role in order to explain the mean free paths of cosmic rays, which is also consistent with the Cluster measurements in the solar wind.

Interstellar Medium

Interstellar medium (ISM) can been recognized as allegoric shapes of nebulae. Some parts of ISM are fully ionized gas due to low density and high temperature and the application of plasma physics is valid. Other parts of ISM consist of neutral gases (atomic or molecular hydrogen, and other molecules) due to high density and low temperature. The existence of turbulence in interstellar medium has been suggested by the power-law curve in the electron density spectrum and as well as the scaling of velocity dispersion with respect to the spatial size.

The spectrum of electron density fluctuation in the interstellar medium has been obtained by different methods and in various spectral ranges from 10^{-4} AU to 10^2 pc [16, 17]. It is found to follow a three-dimensional Kolmogorov law, $k^{-11/3}$, which yields the spectrum $k^{-5/3}$ when integrated with solid angles in the wave vector domain (by multiplying by $4\pi k^2$). Montgomery et al. found that this spectral curve can be explained by the treatment of weakly compressible MHD turbulence [18].

The scaling for velocity dispersion has been obtained by the investigation of spectral line widths from dense, molecular clouds. It was found that the line widths are much broader than that expected from the (low) gas temperature at about 10 K [19, 20]. Though the importance of plasma turbulence in such molecular clouds is not known, the broad velocity dispersion suggests that turbulence in molecular clouds must be of supersonic nature. Whether supersonic random motions can be maintained in turbulence or if such turbulence decays quickly due to shock formation at every fluctuation motion is not yet fully answered. Nonetheless, the observations exhibit that the scaling law can be fitted by the relation $\delta v \sim l^\alpha$ with $\alpha \simeq 0.38$ on scales from 0.1 to 100 pc, and this exponent is close to the Kolmogorov value, $\alpha = 1/3$. Though Kolmogorov's theory refers to incompressible turbulence (hence subsonic turbulence), numerical simulations suggest that the scaling law is valid for supersonic turbulence, too [21].

Accretion Disks

Accretion disks are a widespread phenomenon in astrophysics, and plasma turbulence is essential for its existence. Accretion disks can occur on various sizes, but there are roughly three groups of accretion disks: (1) proto-stellar disks, (2) binary system, and (3) active galactic nuclei. The central objects can be proto-stars, stars, or compact objects like neutron stars or black holes. Accretion refers to accumulation of mass (plasmas, neutral gases, and dusts) onto a central object due to gravitational attraction. In addition, the in-falling mass and the central object have rotation such that the in-falling mass experiences the centrifugal force. As a result, accumulation occurs in the direction of the rotation axis, and a disk is formed around the axis. The system is roughly in equilibrium such that the radial component of gravitational force is balanced by the centrifugal force, and the axial component by the pressure gradient (hydrostatic equilibrium).

The disk material moves almost on Keplerian orbits with decreasing angular velocity in the radial direction, $\Omega(r) \propto r^{-3/2}$. The angular momentum is a product of radial distance and momentum, and is proportional to $r^2\Omega \propto r^{1/2}$. This means that when the matter moves or is transported inward, the excess in angular momentum must be transported outward (due to conservation of total angular momentum). The rate of mass accretion is in fact determined by the radial transport process of angular momentum. However, it was demonstrated that disks are hydrodynamically stable in the sense that perturbation of Keplerian disk leads to epicyclic motion and the gas does not fall onto the central object [22]. Balbus and Hawley [23], on the other hand, showed that the disks are unstable when treated with magnetohydrodynamics, and that the transport of angular momentum can be achieved. Accretion disk dynamics is significantly different between hydrodynamic and magnetohydrodynamic treatments. The disk instability mechanism by Balbus and Hawley is referred to as the magneto-rotational instability. It is not clear if the picture of plasma turbulence in the solar wind is applicable to accretion disks, since the effect of rapid rotation must be taken into account and furthermore, the sense of transport of mass and angular momentum is opposite in accretion disks. Still, the magneto-rotational instability appears to satisfy most of the requirements for an adequate transport mechanism in accretion disks.

5.3 Space Weather and Earth Science

5.3.1 Geomagnetic Storms and Substorms

Disturbance of geomagnetic field is triggered by a change in the solar wind. In particular, two mechanisms are known for the cause of geomagnetic field disturbances. One is the hit of coronal mass ejection at the Earth magnetosphere. Coronal mass ejection, or CME in abbreviation, is an explosion event at the solar surface (most likely caused by the magnetic reconnection process), and releases a huge amount of energy (plasma and magnetic field) on a relatively short time scale of order of hours. The magnetized plasma cloud flows and expands in the interplanetary space at speeds about 1000 km. Not all CMEs encounter the Earth magnetosphere, but when it hits, the CME causes a geomagnetic storm, making the Earth magnetosphere shrink due to the high dynamic pressure.

The other mechanisms is the reconnection process between the interplanetary magnetic field and the Earth dayside magnetic field as illustrated in Fig. 5.4. This happens when the interplanetary magnetic field has a southward component, opposite sense to the Earth dayside magnetic field (pointing from south to north). The reconnected field line is transported by the solar wind to the tail of magnetosphere, enhancing the anti-parallel field line configuration in the tail. This makes the magnetotail unstable, and releases plasma jets by the reconnection process. The jets are aligned with the tail direction and have a diverging pattern: one jet is toward the Earth from the tail; and the other is away from the Earth. Reconnection in the magnetotail (after the dayside reconnection) causes energy transport from the tail to the Earth ionosphere, creating geomagnetic field disturbance (referred to as the substorm). This circulation pattern was proposed by Dungey [24] as illustrated in Fig. 5.4. Many observations suggest that the substorm triggered by the dayside reconnection happens even the solar wind is quiet (i.e., no CME or only small disturbance). The critical point for the onset of the substorm is the direction of the interplanetary magnetic field, i.e., whether or not it is in the parallel sense or anti-parallel with respect to the Earth dayside magnetic field direction. There are various factors determining the southward component in the interplanetary magnetic field: coronal magnetic field structure that is transported by the solar wind; solar wind turbulence;

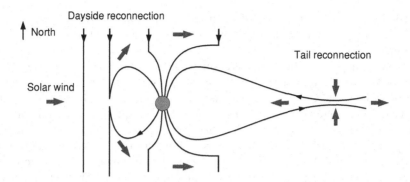

Fig. 5.4 Dungey's notion on magnetic reconnection causing geomagnetic substorm under the southward interplanetary magnetic field

shock waves such as CMEs. Reconnection and turbulence are in close relation with plasma physics. Reconnection is one of drivers of turbulence, but reconnection can be triggered in turbulent fluctuations. Under what conditions reconnection occurs is one of the unsolved problems in plasma physics.

Dynamics of the Earth magnetosphere or in near-Earth space has impacts on our daily life and space engineering, and this field of research is referred to as the space weather. For example, operation of electric devices and facilities in the polar region is subject to the changes in the solar wind and the Earth magnetosphere. Ions and electrons are accelerated during the geomagnetic storms and substorms in the magnetosphere. They stream along the magnetic field and fall onto the polar ionosphere and atmosphere. Aurora is one of the visible phenomena of the interaction between the solar wind and the Earth magnetosphere. It is caused by electron precipitation with energy about several keV coming from the magnetosphere onto the ionosphere, and represents emission primarily from atomic oxygen at visible, discrete wavelengths, forming filamentary structures. The magnetosphere and the ionosphere are electrically and magnetically coupled to each other by magnetic field and electric current (carried by ions and electrons). The current flows not only vertically in the sense of coupling between the magnetosphere and ionosphere, but also horizontally in the ionosphere. Time variation of the ionospheric current has immediate impacts on the Earth ground, causing the current on the ground by induction. Electrically conducting materials are thus subject to induction due to the ionospheric current, e.g., power grids, railways, and pipelines in the polar regions. The current in the ionosphere can reach hundred thousands of Ampères, and this burdens transformers on the power grid and also causes corrosion of pipelines.

There are many other phenomena and effects in space weather. Energetic particles from Earth's magnetosphere, the Sun, and the galaxy can be hazardous to spacecraft at solar panels and on-board electronics. Radiation doze is also higher for astronauts, airline passengers, and crews due to the energetic particles. The Earth atmosphere becomes heated by the radiation from solar flares, and expands to higher altitudes. This causes enhanced airdrag at spacecraft, and can change the orbit and the orientation of the spacecraft. Enhanced ionization degree in the ionosphere due to solar flares and energetic particles can disturb signals from navigation and telecommunication satellites.

There are in principle two methods to study the space weather for forecasting. One is the observations of the Sun and the interplanetary space using remote-sensing. Solar eruption (flares and CMEs), sunspots, and the configuration of the solar magnetic field can be measured using telescopes on board in the spacecraft (to avoid disturbance by the Earth atmosphere). The other method is in situ measurements of the solar wind plasma and magnetic field in front of Earth's magnetosphere. The north-south component of the interplanetary magnetic field is one of the most important parameters determining the dynamics of Earth's magnetosphere. These two methods have close relations with the observation of plasma turbulence at the Sun and in the solar wind.

5.3.2 Modulation of Cosmic Ray Flux

While changes in the solar or interplanetary magnetic field cause geomagnetic storms and substorms at Earth, the field protects us from the galactic cosmic ray. It is known that the flux of galactic cosmic ray is anti-correlated with the solar magnetic activity. Figure 5.5 displays the anti-correlation in these two quantities for the last 60 years. The upper panel is the variation of cosmic ray flux measured by the neutron monitor at Climax, Colorado. Neutrons are generated by the hit of cosmic ray at the Earth atmosphere. The cosmic ray flux is not constant but oscillates with the relative amplitude about 15 % to the first order with 11-year cycle (from one maximum to another maximum), and to the second order with 22-year cycle (one maximum has a flatter curve, and the other maximum has a pointed curve). It should be noted that the variation amplitude of the cosmic ray flux is much larger than that of radiation energy coming from the Sun. The intensity of solar radiation reaching the Earth has been relatively constant through the last 2000 years, with variations of around 0.1–0.2%. The lower panel is the variation of the sunspot numbers. The 11-year cycle of the solar activity is clearly visible in the data. The cosmic ray data and the sunspot data are correlated to each other, in the sense of anti-correlation that the maximum of the cosmic ray flux corresponds to the minimum of the sunspot numbers. Furthermore, the 22-year cycle in the cosmic ray data is suggestive of correlation with the polarity change of the solar magnetic field.

Modulation of cosmic ray flux with periods 11 years and 22 years can be explained primarily by the geometry of interplanetary magnetic field [25]. Cosmic ray mostly consist of protons and their trajectories are subject to the magnetic field structure and solar eruption events (CMEs) in the heliosphere. For example, during the active phase of solar magnetic field (i.e., many sunspots) the shielding of cosmic ray flux is more effective because CMEs frequently occur and they scatter the cosmic ray particles. This creates the anti-correlated variations with 11-year cycle between the neutron monitor counts and the sunspot numbers. Furthermore, cosmic ray particles in the heliosphere stream in different fashions, depending on the phase of solar magnetic field configuration. When the solar magnetic polarity is positive (the field line goes from north to south outside the Sun), cosmic ray trajectories are statistically in the direction from heliospheric poles (north and south) to the solar equatorial plane (roughly close to the Earth orbital plane). When the polarity of the solar magnetic field is negative (or the field line is from south to north) the particle trajectories are in the opposite direction, from the equatorial plane to the heliospheric pole directions. The neutral lines of the solar magnetic field in the interplanetary space form a curved plane or surface with wavy structure, and the combination of the magnetic-neutral plane and the polarity change of the Sun has a shielding effect with 22-year cycle against cosmic ray flux to the heliosphere and to the Earth. While cosmic ray modulation can be explained to the first order by the solar cycle and the polarity reversal of the solar magnetic field, it is also influenced by turbulence in the solar wind as is the case for cosmic ray transport in the interstellar space. How the turbulence properties and the fluctuation geometry evolves together with the solar cycle is an important subject to study.

What kind of consequences do the solar activity and the cosmic ray modulation have at the Earth? The sunspot number data have been recorded since the early seventeenth century (ca. 1610–1611) until today. The sunspot number, as displayed in Fig. 5.5, oscillates with the period of approximately 11 years, but the amplitude (maximum sunspot number) is not the same from one solar cycle to another. Rather, the solar cycle itself exhibits variation on large time scales. Furthermore, there was a period of only few sunspots for many decades from 1645 to 1715, called the Maunder minimum [26]. In the late twentieth century the sunspot number reaches typically 40,000 to 50,000 spots, whereas during the Maunder minimum only about 50 sunspots were observed for 30 years. The sunspot number data also exhibit another minimum phase, the Dalton minimum around from about 1790 to 1830.

The Maunder minimum happened to be the coldest phase of the Little Ice Age. During that period the climate in Europe and North America was characterized by cold winters. The question whether or not there is causality between the solar activity (variation in sunspots) and the Earth climate has not been proved or answered, but still it raises a hypothesis that past climate changes in the Earth were influenced by variations in solar activity in the sense that lower temperatures in the Earth is correlated with lower solar activity or fewer sunspot numbers rather than correlation with radiation from the Sun. Variation in the solar radiation is of the order 0.1%, while variation in cosmic ray modulation via changes in the solar magnetic field reaches about 15%. For example, cosmic ray may promote condensation process of water molecule in the atmosphere by the same principle as the cloud chamber experiment, such that enhanced cosmic ray flux results in enhanced cloud formation. One has to bear in mind that the idea is still speculative; data of cosmic ray flux are limited to only few solar cycles; relation between the cloud formation and the global temperature in the atmosphere is not clear. Moreover, one has to take account of terrestrial factors such as continental drift (plate tectonics), volcanic activity, and anthroporogenic factors.

Fig. 5.5 Neutron monitor
counts (counts per hour
divided by 100) and sunspot
numbers for solar cycles 19
through 23. The neutron
monitor data are provided by
the University of New
Hampshire. National Science
Foundation Grant
ATM-0339527 is
acknowledged. The sunspot
number data are provided by
World Data Service for the
Sunspot index, SIDC, Royal
Observatory of Belgium,
online catalog of the sunspot
index:http://www.sidc.be/
sunspot-data/

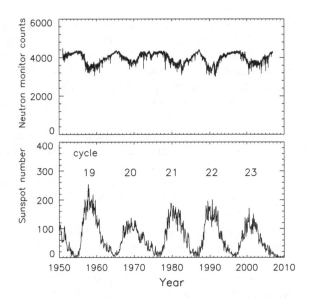

While the data of sunspot numbers are available since the last 400 years, there
are indirect methods, proxy, using isotopes such as carbon-14 and beryllium-10
that are generated mainly by cosmic ray hit at the atmosphere to investigate the
relation between solar activity and the Earth climate on longer time scales in the
past [27, 28]. One example is the study of isotopes carbon-14 and and oxygen-18
[29]. Carbon-14 is created by the cosmic ray hit at the Earth atmosphere, and can
be used as proxy of cosmic ray flux. It is stored in trees by uptake process, and
its time profile can be determined using the method of dendrochronology (profile
in tree ring evolution). The carbon-14 data exhibit quasi-periodic variations on the
time scale 6,000 to 10,000 years. This can be compared with the profile of oxygen-
18 which is a stable isotope and the ratio to the oxygen-16 is used as proxy for
rainfall. (The ratio is different between air and ocean.) The profile of oxygen-18
obtained from stalagmites and calibrated against radioactive decay of uranium and
thorium also exhibits quasi-periodic variations on the same time scale. Again, one
has to be careful (i.e., correlation does not mean causality), but the study is one of
demonstrations that the cosmic ray variation is quasi-periodic on large time scale, and
furthermore there is an evidence of anti-correlation between cosmic ray modulation
and rainfall amount. One of hypothetical explanations is that cosmic ray hit at the
atmosphere promotes cloud formation due to condensation of water molecules in
favor of anti-correlation that intense cosmic ray flux results in larger coverage of
cloud and possibly resulting in temperature variation in the atmosphere [30, 31].

How variations in solar activity influence the Earth is one of highly interdisciplinary research fields. The Sun provides the Earth not only with light and heat but also
with the shielding effect from galactic cosmic ray in the heliosphere. The relation
between the Sun and the Earth can be found on various time scales. On a short time

scale the solar activity influences the space weather, and on a large time scale it might influence the climate system in the Earth.

References

1. Biskamp, D.: Magnetohydrodynamic Turbulence. Cambridge University Press, Cambridge (2003)
2. Longtin, M., Sonnerup, B.: Modulational instability of circularly polarized Alfvén waves. J. Geophys. Res **91**, 798–801 (1986)
3. Terasawa, T., Hoshino, M., Sakai, J.-.I., Hada, T.: Decay instability of finite-amplitude circularly polarized Alfvén waves: A numerical simulation of stimulated brillouin scattering. J. Geophys. Res. **91**, 4171–4187 (1986)
4. Wong, H.K., Goldstein, M.L.: Parametric instabilities of circularly polarized Alfvén waves including dispersion. J. Geophys. Res. **91**, 5617–5628 (1986)
5. Davidson, P.A.: Turbulence: an introduction for scientists and engineers. Oxford University Press, New York (2004)
6. Bale, S.D., Kellogg, P.J., Mozer, F.S., Horbury, T.S., Reme, H.: Measurement of the electric fluctuation spectrum of magnetohydrodynamic turbulence. Phys. Rev. Lett. **94**, 215002 (2005). doi: 10.1103/PhysRevLett.94.215002
7. McKenzie, J.F., Westphal, K.O.: Transmission of Alfvén waves through the Earth's bow shock. Planet. Space Sci. **17**, 1029–1037 (1969)
8. McKenzie, J.F., Westphal, K.O.: Interaction of hydromagnetic waves with hydromagnetic shocks. Phys. Fluids. **13**, 630–640 (1970)
9. McKenzie, J.F.: Hydromagnetic wave interaction with the magnetopause and the bow shock. Planet. Space Sci. **18**, 1–23 (1970)
10. Gary, S.P.: Theory of Space Plasma Microinstabilities. Cambridge Atmospheric Space Science Series, Cambridge (1993)
11. Narita, Y., Glassmeier, K.-H., Goldstein, M.L., Motschmann, U., Sahraoui, F.: Three-dimensional spatial structures of solar wind turbulence from 10,000-km to 100-km scales, Ann. Geophys. **29**, 1731–1738 (2011)
12. Swordy, S.P.: The energy spectra and anisotropies of cosmic rays. Space Sci. Rev. **99**, 85–94 (2001)
13. Drury, L.Ò.C.: An introduction to the theory of diffusive shock acceleration of energetic particles in tenuous plasmas. Rep. Prog. Phys. **46**, 973–1027 (1983)
14. Bieber, J.W., Matthaeus, W.H., Smith, C.W., Wanner, W., Kallenrode, M.-.B., Wibberenz, G.: Proton and electron mean free paths: The Palmer consensus revisited. Astrophys. J. **420**, 294–306 (1994)
15. Bieber, J.W., Wannger, W., Matthaeus, W.H.: Dominant two-dimensional solar wind turbulence with implications for cosmic ray transport. J. Geophys. Res. **101**, 2511–2522 (1996)
16. Armstrong, J.W., Cordes, J.M., Rickett, B.J.: Density power spectrum in the local interstellar medium. Nature **291**, 561–564 (1981)
17. Armstron, J.W., Ricket, B.J., Spangler, S.R.: Electron density power spectrum in the local interstellar medium. Astrophys. J. **443**, 209–221 (1995)
18. Montgomery, D., Brown, M.R., Matthaeus, W.H.: Density fluctuation spectra in magnetohydrodynamic turbulence. J. Geophys. Res. **92**, 282–284 (1987)
19. Larson, R.B.: Turbulence and star formation in molecular clouds. Mon. Notes Roy. Astron. Soc. **194**, 809–826 (1981)
20. Falgarone, E., Puget, J.-.L., Pérault, M.: The small-scale density and velocity structure of quiescent molecular clouds. Astron. Astrophys. **257**, 715–730 (1992)

21. Porter, D.H., Pouquet, A., Woodward, P.R.: Kolmogorov-like spectra in decaying three-dimensional supersonice flows. Phys. Fluids **6**, 2133–2142 (1994)
22. Balbus, S.A., Hawley, J.F., Stone, J.M.: Nonlinear stability hydrodynamical turbulence and transport in disks. Astrophys. J. **467**, 76–86 (1996)
23. Balbus, SA., Hawley, J.F.: A powerful local shear instability in weakly magnetized disks. Astrophys. J. **376**, 214–222 (1991)
24. Dungey J.W. 1961 Interplanetary magnetic fields and the auroral zones. Phys. Rev. Lett. **6**, 47–48
25. Kota, J., Julipii, J.R.: 3-D modeling of cosmic-ray transport in the heliosphere: Toward solar maxium. Adv. Space Res. **27**, 529–534 (2001)
26. Eddy, J.A.: The Maunder Minimum. Science **192**, 1189–1202 (1976)
27. Stuiver, M., Braziunas, T.F.: Atmospheric ^{14}C and century-scale solar oscillations. Nature. **338**, 405–408 (1989)
28. Hoyt, D.V., Schatten, K.H.: The Role of the Sun in Climate Change. Oxford University Press, Oxford (1997)
29. Neff, U., Burns, S.J., Mangini, A., Mudelsee, M., Fleitmann, D., Matter, A.:. Strong coherence between solar variability and the monsoon in Oman between 9 and 6 kyr ago. Nature **411**, 290–293 (2001)
30. Svensmark, H., Friis-Christensen, E.: Variation of cosmic ray flux and global cloud coverage: A missing link in solar-climate relation- ships. J. Atmos. Sol. Terr. Phys. **59**, 1225–1232 (1997)
31. Svensmark, H.: Cosmoclimatology: A new theory emerges. Astron. Geophys. **48**, 18–24 (2007)

Index

Y. Narita, *Plasma Turbulence in the Solar System*, SpringerBriefs in Physics,   101
DOI: 10.1007/978-3-642-25667-7, © The Author(s) 2012